Basic
Tig & Mig Welding
(GTAW & GMAW)

THIRD EDITION

Basic Tig & Mig Welding
(GTAW & GMAW)

IVAN H. GRIFFIN • EDWARD M. RODEN •
CHARLES W. BRIGGS

DELMAR PUBLISHERS INC.

Cover photo by Larry Jeffus

Administrative editor: Mark Huth
Production editor: Eleanor Isenhart

For information address Delmar Publishers Inc.
2 Computer Drive West, Box 15-015
Albany, New York 12212

Printed in the United States of America
Published simultaneously in Canada
by Nelson Canada,
a Division of International Thomson Limited.

10 9 8 7 6 5 4 3

Library of Congress Cataloging in Publication Data

Griffin, Ivan H.
   Basic TIG & MIG welding.

   Includes index.
   1. Gas tungsten arc welding.   2. Gas metal arc
welding. I. Roden, Edward M.   II. Briggs, Charles W.
III. Title.   IV. Title: Basic T.I.G. & M.I.G. welding.
V. Title: Basic TIG and MIG welding.   VI. Title:
Basic T.I.G. & M.I.G. welding.
TK4660.G744   1984   671.5'212   83-18844
ISBN 0-8273-2129-5

CONTENTS

CHARTS

PREFACE

Inert gas welding refers to welding processes in which an electric arc is completely shielded by a chemically inert gas. The two principal inert gas processes are gas tungsten arc welding, GTAW, and gas metal arc welding, GMAW. Although these processes are not new to the welding industry, their importance has grown rapidly in recent years. *Basic TIG and MIG Welding* (GTAW and GMAW) helps the beginning welder develop skill in these important processes.

This basic textbook provides the beginning welder with an opportunity to develop skills by following step-by-step procedures. The text is divided into two sections. In the first section, TIG Welding, the beginning welder is introduced to the principles and equipment used in the gas tungsten arc welding process. Other units in section one include step-by-step procedures for welding both steel and nonferrous metals. In the second section, MIG Welding, the beginning welder is introduced to the MIG process and equipment used in MIG welding. This section also includes specific procedures for MIG welding on both steel and nonferrous metals.

This revision of *Basic TIG and MIG Welding* includes photographs of welds in progress. The reader can see what takes place in the weld zone. No previous skill, knowledge, or training in welding is required for the student or instructor to use this textbook. All of the unit material, both new and old has been reviewed for readability and reliability. The addition of new illustrations helps to update the text. An index has been added to give the student fast and easy reference.

Because of its clear and readable format, this text has proved popular in prevocational, industrial arts, adult education and occupational education programs. The text is used extensively for training welders, auto body mechanics, machinists, and a variety of other vocational areas.

SECTION 1

TIG welding (GTAW)

TIG (or GTAW — gas tungsten-arc welding) is basically a form of arc welding. It is especially useful in welding aluminum. Developed in the period of 1940 to 1960, it has rapidly become one of the indispensable welding methods.

The equipment used is more complex and expensive than that used for arc welding because electricity, water and gas must all be provided and controlled. MIG (metal inert-gas) welding is closely related to TIG welding.

THE TUNGSTEN INERT-GAS
SHIELDED-ARC WELDING PROCESS

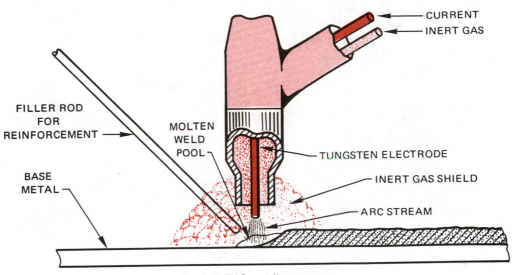

CURRENT
INERT GAS

FILLER ROD
FOR
REINFORCEMENT

MOLTEN
WELD
POOL

TUNGSTEN ELECTRODE

INERT GAS SHIELD

BASE
METAL

ARC STREAM

Fig. 1-1 TIG welding process

The tungsten inert-gas shielded-arc welding process, figure 1-1, is an extension, refinement, and improvement of the basic electric arc welding process.

In the complete name of this process:

Tungsten refers to the nonconsumable electrode which conducts electric current to the arc.

Inert refers to a gas which will not combine chemically with other elements.

Gas refers to the material which blankets the molten puddle and arc.

Shielded describes the action of the gas in excluding the air from the area surrounding the weld.

Arc indicates that the welding is done by an electric arc rather than by the combustion of a gas.

The process is commonly referred to as *TIG welding* (or GTAW — gas tungsten-arc welding) which is obtained from the first letter of each of the words, tungsten, inert, and gas. This type of welding is often referred to as Heliarc®, which is the trade name of a particular manufacturer. The TIG process generally produces welds which are far superior to those made by other welding processes.

ELEMENTS OF THE PROCESS

As shown in figure 1-2, the basic process uses an intense arc drawn between the work and a tungsten electrode. The arc, the electrode, and the weld zone are surrounded by an inert gas which displaces the air to eliminate the possibility of contamination of the weld by

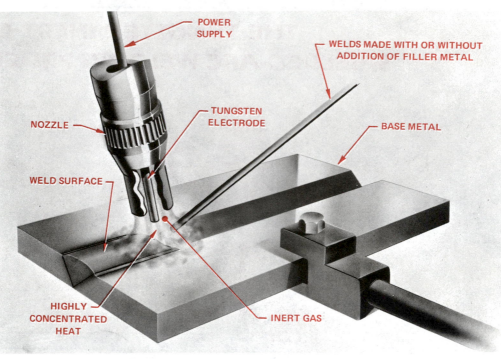

Fig. 1-2 Essentials of the TIG welding process

oxygen and nitrogen in the atmosphere. The tungsten electrode has a very high melting point (6,900 degrees F.) and is almost totally nonconsumable when used within the limits of its current-carrying capacity.

The inert gas supplied to the weld zone is usually either helium or argon, neither of which will combine with other elements to form chemical compounds. Argon gas is usually recommended because it is more generally available and better suited for use in the welding of a wide variety of metals and alloys. The basic components for a water-cooled TIG welding outfit are indicated in figure 1-3.

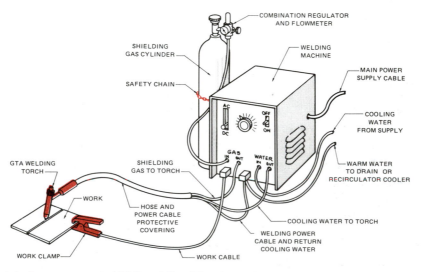

Fig. 1-3 Essentials for water-cooled TIG welding (Reprinted from Jeffus & Johnson, *Welding: Principles & Applications,* Figure 1-6. © 1984 by Delmar Publishers Inc.)

ADVANTAGES OF TIG WELDING

Examples of the beads welded by arc, oxyacetylene and TIG processes are shown in figure 1-4.

- No flux is required and finished welds do not have to be cleaned of corrosive residue. The flow of inert gas keeps air away from the molten metal and prevents contamination by oxygen and nitrogen.

- In the chemical composition, the weld itself is usually equal to the base metal being welded. It is usually stronger, more resistant to corrosion, and more *ductile* (ability of a metal to deform without fracturing) than welds made by other processes. The inert gas will not combine with other elements or permit contamination by such elements, thus keeping the metal pure.

- Welding can be easily done in all positions. There is no *slag* (waste material entrapped in weld) to be worked out of the weld.

- The welding process can be easily observed. No smoke or fumes are present to block vision, and the welding puddle is clean.

- There is minimum distortion of the metal near the weld. The heat is concentrated in a small area resulting in a small heat-affected zone.

- There is no splatter to cause metal-cleaning problems. Since no metal is transferred across the arc, this problem is avoided.

ARC WELDING

OXYACETYLENE WELDING

TIG WELDING

Fig. 1-4 Comparison of beads as welded

- Practically all the metals and alloys used industrially can be fusion welded by the TIG process in a wide variety of thicknesses and types of joints.

- TIG welding is used particularly for aluminum and its alloys (even in very thick sections), magnesium and its weldable alloys, stainless steel, nickel and nickel-base alloys, copper and copper alloys, some brasses, low alloy and plain carbon steel, and the application of hard-facing alloys to steel.

REVIEW QUESTIONS

1. What does the term TIG welding refer to?

2. What are the essentials of TIG welding?

3. How does a TIG weld compare chemically with a metal arc weld?

4. How do the mechanical properties of TIG welds compare with those of welds made by other manual processes and with the base metal?

5. What is meant by the term slag entrapment? Why is it harmful?

6. How do the uses of TIG welding compare with other manual processes?

7. Why is argon recommended as the shielding gas for most TIG welding?

EQUIPMENT FOR MANUAL TIG WELDING

The equipment and material required for TIG welding consist of an electrode holder, or torch, containing gas passages and a nozzle for directing the shielding gas around the arc; nonconsumable tungsten electrodes; a supply of shielding gas; a pressure-reducing regulator and flowmeter; an electric power unit; and on some machines a supply of cooling water.

THE TORCH

A specially designed torch is used for TIG welding. It is so constructed that various sizes of tungsten electrodes can be easily interchanged and adjusted. The torch is equipped with a series of interchangeable gas cups to direct the flow of the shielding gas. Some of the torches are air-cooled, but water-cooled torches are more widely used.

SOURCE OF ELECTRIC CURRENT

The source of the electric current used in modern TIG welding is a specially designed welding machine, figures 2-1 and 2-2. It is possible to adapt standard alternating-current (AC) and direct-current (DC) welding machines such as are used in arc welding operations to TIG welding operations. However, a unit of this type is bulky and hard to manage when compared with the modern machines that are designed for TIG welding. An AC arc is best

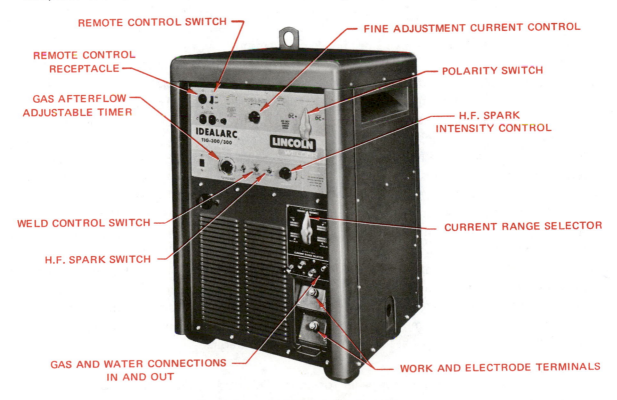

Fig. 2-1 AC - DC TIG welding machine

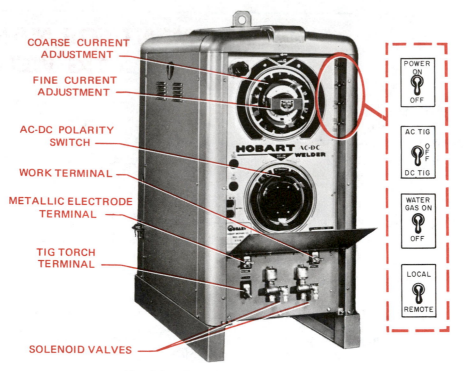

COARSE CURRENT ADJUSTMENT

FINE CURRENT ADJUSTMENT

AC-DC POLARITY SWITCH

WORK TERMINAL

METALLIC ELECTRODE TERMINAL

TIG TORCH TERMINAL

SOLENOID VALVES

POWER ON OFF

AC TIG OFF DC TIG

WATER GAS ON OFF

LOCAL REMOTE

Fig. 2-2 AC - DC TIG welding machine

suited for aluminum and some other metals and alloys. But a standard 60-cycle alternating current, which changes its direction of flow 120 times a second, is unsuited for welding because the electrical characteristics of the oxides on these metals cause the arc to extinguish (go out) at every half cycle or change of direction.

If, however, an *igniter arc current* is added to the standard 60-cycle current, the tendency to extinguish will be overcome because the igniter current will maintain a path for the standard 60-cycle current to follow. This igniter current is usually generated within the machine by a spark-gap oscillator which causes the current to change direction, not 120 times a second, but millions of times each second. Because this frequency of change is very high, the term high frequency is used in describing this current. Since standard 60-cycle alternating current actually does the welding, it is called AC welding; hence the term, *high-frequency alternating current* welding or, as it is commonly referred to, HFAC.

TIG welding power units, in most cases, can supply either AC or DC power to the electrode. These welding machines are equipped with a high-frequency oscillator which injects the high-frequency igniter current into the welding circuit. This high-frequency current causes a spark to jump from the electrode to the work without contact between the two. This ionizes the gap and allows the welding current to flow across the arc. Some manufacturers provide an external means of varying the frequency of this alternating current. In other machines the housing must be opened to make adjustments. High frequency stabilizes the arc. Arc stability is essential for welding thin-gage materials.

CONTROLS

Figure 2-3 shows the control panel for an AC-DC welding machine. TIG power units are usually equipped with solenoid valves to turn the flow of shielding gas and cooling water

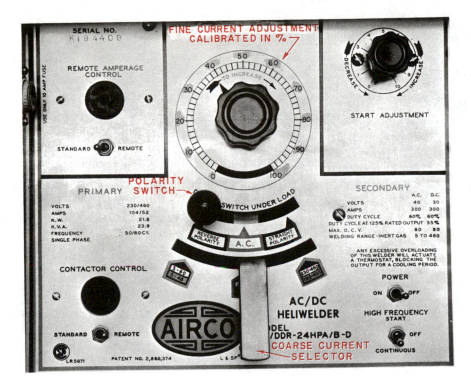

Fig. 2-3 Control panel for AC-DC welding machine

on and off. They are also provided with a remote-control switch, either hand- or foot-operated, to turn the water and gas on and off. Some of these remote-control devices also turn the main welding current on or off at the same time.

Most manufacturers equip the solenoid valves with a delayed-action device which allows the cooling water and shielding gas to continue to flow after the remote-control switch has been set at the stop or open position. This delay allows the tungsten electrode to cool to the point that it will not oxidize when the air comes in contact with it.

Note: Oxide is a scale that forms on a metal when it is exposed to air, and especially when it is heated.

Some machines have an external means of varying the time of this afterflow to correspond to the electrode which is being used. In other types of machines, the housing must be opened to make this adjustment, figure 2-4. In any case, the shielding gas must be allowed to flow long enough so that the tungsten electrode cools until it has a bright, shiny surface.

SHIELDING GAS

The shielding gases are distributed in standard cylinders, which contain 330 cubic feet at 3,000 p.s.i.

As with all compressed gases, a regulator must be provided to reduce the high cylinder pressure to a safe, usable working pressure.

The main difference between the regulators used for oxyacetylene welding and those used for TIG welding is that the working pressure on the oxyacetylene regulators is indicated in pounds per square inch while the regulators used for TIG welding indicate the flow of

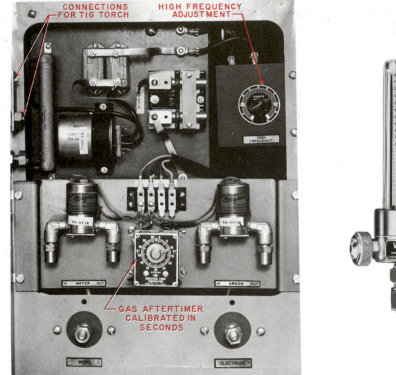

Fig. 2-4 Lower portion of machine: cover removed

Fig. 2-5 Combination regulator and flowmeter

shielding gas in cubic feet per hour. The latter are generally referred to as *flowmeters*. A combination regulator and flowmeter is shown in figure 2-5.

Another significant difference between a standard regulator and the flowmeter is that the regulator will indicate the working pressure to the torch regardless of the regulator's position, while the tube on the flowmeter must be in a vertical position if an accurate reading is to be obtained.

REVIEW QUESTIONS

1. How long should the shielding gas and cooling water be allowed to flow after the welding arc is broken?

2. How do modern TIG welding machines compare with earlier models?

3. What makes the modern TIG torch adaptable to a wide range of welding operations?

4. What precaution is used to install a flowmeter on a gas cylinder?

5. In the air-cooled TIG torch, what does the cooling?

THE WATER-COOLED TIG WELDING TORCH

The TIG torch, figure 3-1, is a multipurpose tool. It serves as:

- A handle.
- An electrode holder.
- A means of conveying shielding gas to the arc.
- A conductor of electricity to the arc.
- A method of carrying cooling water to the torch head.

COOLING WATER FLOW

Figure 3-2 shows a cross section of a torch with the cooling water flow indicated. The water cools the torch head, the collet and the electrode; it also cools the relatively light welding current cable which will overheat and burn if it is not surrounded by cooling water at

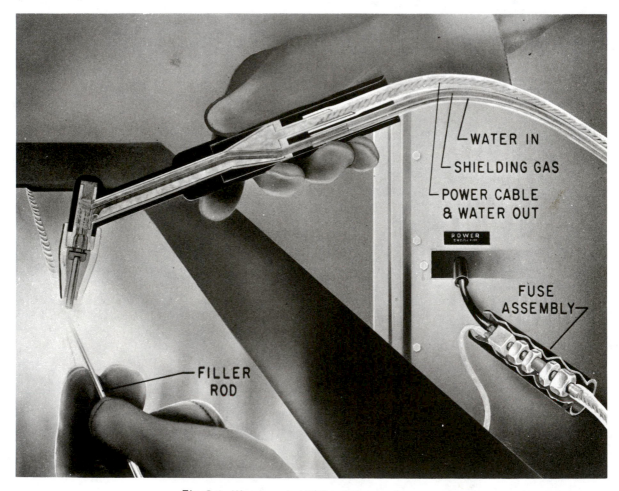

Fig. 3-1 Water-cooled TIG welding torch

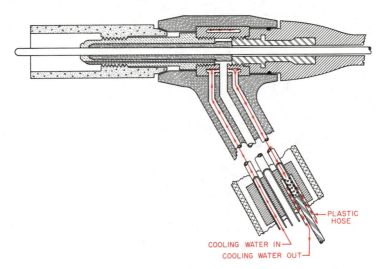

Fig. 3-2 Cooling water circuit

all times when current is flowing. All equipment manufacturers will supply cooling water requirements. A typical recommendation for cooling a medium-duty (300-amp.) torch is one quart of water per minute at 75 degrees F. (24 degrees C) or less, at not over 50 pounds per square inch pressure.

THE FLOW OF GAS

Figure 3-3 indicates the path of the regulated shielding gas through a hose to the torch head and through the collet holder. The gas then flows through a series of holes around the collet holder which direct the flow around the tungsten electrode through the ceramic nozzle to the work zone. The diameter and length of this nozzle vary with the size of the electrode used, the type of current being used, the material being welded, and the shielding gas being used.

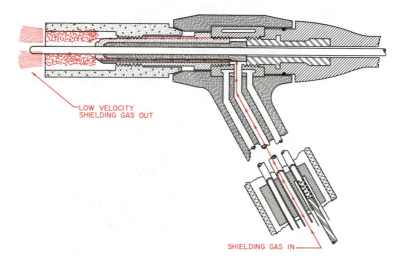

Fig. 3-3 Shielding gas flow

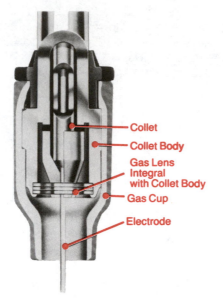

Collet

Collet Body

Gas Lens
Integral
with Collet Body

Gas Cup

Electrode

Fig. 3-4 Gas lens

Fig. 3-5 Effect of gas lens

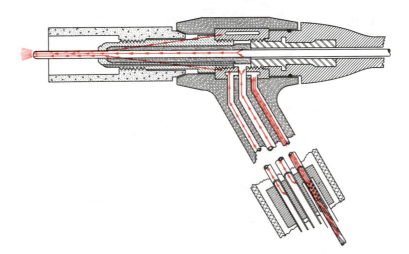

Fig. 3-6 Electric circuit

Note: A gas lens, figure 3-4, improves the stability of the shielding gas. A torch equipped with a gas lens will project the shielding gas to a greater distance, figure 3-5. The nozzle may be held up to 1 inch above the work. More tungsten can be extended beyond the cup.

THE FLOW OF ELECTRICITY

As shown in figure 3-6, the electric current flows through the water-cooled welding cable, through the torch head to the collet holder and collet, to the tungsten electrode which forms one terminal of the arc, then through the work, and back through the ground cable to the power source. The student should examine a torch and compare the size of this ground cable with that of the cable leading into the torch.

The description of the electric circuit through the torch and work indicates that the electrode is negative or on straight polarity. If reverse polarity is used, the current flow is in the opposite direction.

If 60-cycle alternating current is used, the direction of flow changes 120 times each second so the electrode is positive (+) 60 times each second and negative (–) 60 times each second. This alternating of welding current is found to be very advantageous in many TIG welding operations. Direct-current straight polarity is referred to as DCSP; direct-current reversed polarity, as DCRP; and alternating current as AC, or in the case of AC with a superimposed high-frequency igniter current, as HFAC. The student should learn and remember these abbreviations since they are used in nearly every technical manual and paper written on the subject of TIG welding.

THE ASSEMBLY AND OPERATION OF THE TIG TORCH

To prepare the torch, figure 3-7, for welding operations, first choose the proper size electrode, matching collet and gas cup or nozzle. Some torches also require that the collet holder be changed with each different collet. Figure 3-8 shows a TIG torch with interchangeable collets.

1. Remove the collet cap or gas cap.

2. Remove the nozzle or gas cup by turning it counterclockwise.

3. If the collet holder is removable, take it off by turning it in a counterclockwise direction.

   Note: In the case of a torch with a single collet holder, step 3 can be ignored and the collet and electrode can be removed from the gas cup side of the torch.

4. Remove the collet and the electrode from the collet holder.

5. Choose the proper size electrode, collet holder, collet and gas cup or nozzle.

6. Screw the collet holder firmly into the torch.

7. Screw the nozzle into the collet holder firmly against the O ring on the torch body.

8. Place the proper collet in the collet holder and replace the gas cap or collet cap in the torch, leaving it loose by one-half to one turn.

9. Insert the electrode through the gas nozzle into the collet.

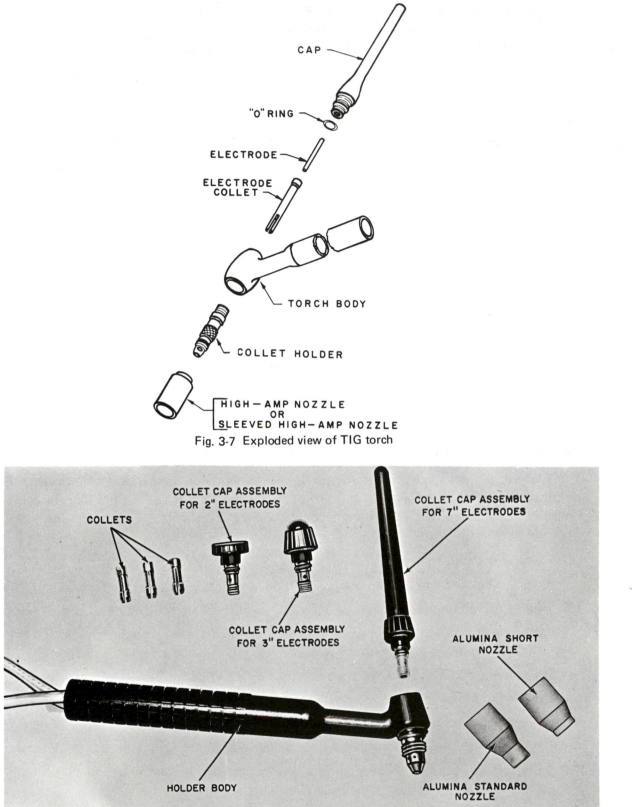

CAP

"O" RING

ELECTRODE

ELECTRODE
COLLET

TORCH BODY

COLLET HOLDER

HIGH—AMP NOZZLE
OR
SLEEVED HIGH—AMP NOZZLE

Fig. 3-7 Exploded view of TIG torch

COLLET CAP ASSEMBLY
FOR 2" ELECTRODES

COLLETS

COLLET CAP ASSEMBLY
FOR 7" ELECTRODES

COLLET CAP ASSEMBLY
FOR 3" ELECTRODES

ALUMINA SHORT
NOZZLE

ALUMINA STANDARD
NOZZLE

HOLDER BODY

Fig. 3-8 TIG torch with interchangeable collets

Note: Never insert the electrode in the collet before inserting the collet in the torch. This guards against inserting a used electrode in the collet and, after use, finding that each end of the electrode has a ball formed on the end thus preventing the removal of the electrode from the collet.

10. Adjust the electrode for the recommended extension beyond the nozzle and tighten the gas cap or collet cap until the electrode is firmly fixed in the torch.

11. When the electrode extension needs to be adjusted to compensate for the slow burn-off, loosen the gas cap and adjust the electrode, and then firmly tighten the cap. Check the electrode to be sure it is firmly seated in the collet.

CAUTIONS:

- The ceramic nozzles are brittle, expensive, and easily broken. Always handle them with great care.

- Any electrical connection that is not thoroughly tight will generate extensive heat and may ruin the torch. Be sure all collet holders, collets and electrodes are tight to avoid costly damage.

- If the nozzle or gas cap is loose, it is possible for the shielding gas to draw air into the torch and contaminate the electrode as well as the weld. Always make sure that these parts are tight and that all O rings are in place.

REVIEW QUESTIONS

1. How does the size of the water-cooled cable to the torch compare with the ground or work cable?

2. What result would be expected if the water-cooled cable were not supplied with the cooling water at all times while the welding operation is going on?

3. What effect will a loose electrode or collet have on the torch?

4. What effect does a loose gas cap or ceramic nozzle have on the electrode and work zone?

5. If the electrode collet is put in the collet holder upside down, what happens?

Unit 4

THE WELDING OF ALUMINUM

Most of the basic research and development in the use of the tungsten inert-gas shielded-arc process has been concentrated on the welding of aluminum and its weldable alloys. A brief analysis of the properties of aluminum will help to account for its widespread popularity.

ADVANTAGES OF ALUMINUM

- It is one of the earth's most abundant metals, making up about eight percent of the earth's crust.
- It has great strength in comparison to its weight.
- It is generally highly resistant to most forms of corrosion.
- It gives a very attractive appearance.
- It is very ductile and malleable.
- It has very good electrical and thermal conductivity characteristics.
- When its other qualities are considered, it is reasonably inexpensive.

As shown, aluminum has many advantages, but certain other facts must be considered before aluminum can be sucessfully joined by any of the welding processes, including TIG.

CONSIDERATIONS

- Chemically, aluminum is a very active metal.
- Aluminum absorbs heat five times faster than steel.
- It combines with oxygen from the atmosphere even at room temperature to form a very hard oxide film on the surface. The hardness is illustrated in abrasive wheels composed of aluminum oxide.
- While aluminum melts at 1218 degrees F., its oxide melts at 3600-3900 degrees F. Even in the molten condition, it has a large amount of surface strength when compared to the oxides of many other metals.
- Aluminum oxide tends to absorb moisture. Under the extreme heat of the welding arc, this moisture breaks down to free hydrogen which often leads to porosity in the weld.
- The oxide can be removed mechanically by filing, scraping, or wirebrushing. It can be removed chemically with some liquid cleaners or by the use of flux. Most important in this discussion, it can be vaporized by the intense heat of the electric arc under the proper circumstances.
- Regardless of how the oxide is removed, it starts to re-form immediately. Too long a time lapse between cleaning and welding leaves the surface in about the same condition as it was before it was originally cleaned.

- Chemical fluxes must be immediately and thoroughly removed to prevent highly corrosive action on the metal. This is no problem in TIG welding because no flux is required.

- Compared to most metals, aluminum has a very high thermal conductivity. Thus the heat is conducted away from the weld zone at a fast rate. This means that a very high heat input must be maintained in the weld zone to balance the heat loss to the adjacent metal. TIG welding, with its intensely hot arc, is an excellent method of maintaining this high heat input.

- Aluminum is easily welded in overhead or other positions by the TIG process. When molten, it has a high surface tension for such a light metal. This surface tension tends to hold the molten puddle in position. The high rate of thermal conductivity causes the molten pool to solidify rapidly.

- Aluminum does not change color when it nears the melting point as do most other metals. When chemical fluxes are used to clean the surface, they cause considerable glare which makes accurate observation of the molten puddle difficult. With TIG welding there is no glare or smoke. The welder has a clear view of the size, shape, and condition of the molten pool at all times.

WELDABILITY OF ALUMINUM ALLOYS

The basic factor governing the ability of an aluminum alloy to be successfully welded is its chemical composition. Whether the alloy is wrought, die cast or sand cast makes little difference in the welding procedure.

According to the Air Reduction Company,

Weldability is the capacity of a metal to be fabricated by welding under the imposed conditions into a structure adequate for the intended purpose.

Figure 4-1 shows a section of aluminum sheet. The alloy, its degree of hardness or temper, and its thickness in thousandths of an inch are indicated.

The chemical composition, physical characteristics and heat treatment of the large array of alloys produced by manufacturers of aluminum and its alloys are out of the scope of this book. However, these manufacturers can supply a large amount of information about their products. A study of material of this type will increase the welder's knowledge and make the welder a more valuable employee.

REVIEW QUESTIONS

1. Would the use of aluminum be expected to increase or decrease in relation to the other commonly used metals? Why?

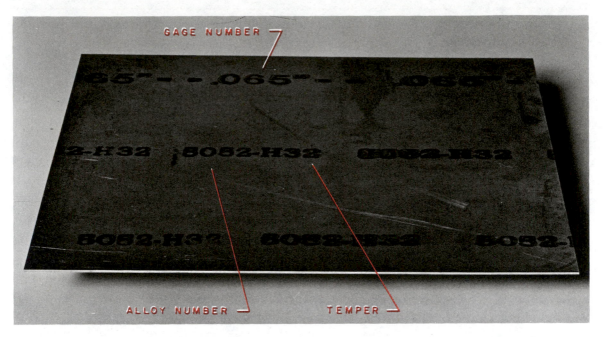

Fig. 4-1 Method of labeling aluminum sheets

2. What is meant by the term thermal and electrical conductivity?

3. Where can information be found about the chemical composition or physical characteristics of the various aluminum alloys?

4. What major condition causes aluminum to be considered difficult to weld?

5. What other characteristics cause aluminum to be considered difficult to weld?

THE ACTION IN AND AROUND THE WELDING ARC

The experiments in this unit will provide an opportunity to investigate some of the phenomena that take place in and around the welding arc. This provides a basis on which to judge the conditions necessary to produce high-quality TIG welds.

Note: As the following experiments are carried out in the unshielded atmosphere, aluminum oxide is formed rapidly. Also, when welding with DCRP the weld zone is covered with black residue. Remember, arc phenomena and not welding techniques are being investigated. Some aluminum has been welded in the past using the carbon arc process DCSP. TIG welding has made this method almost obsolete.

EXPERIMENT 1

Materials

DC welding machine   2 carbon electrodes 1/4-inch diameter

Procedure

1. Grind one end of each electrode to a pencil point.

2. Place one electrode in the holder and the other in a vise or other clamping device which is attached to the ground cable.

3. Set the controls on the machine to produce 150 amps. or more with the electrode holder negative — DCSP.

4. Draw an arc between the pointed ends of the carbon electrode and carbon work. This makes one pole of the arc negative and the other positive, which is a normal condition in all DC welding arcs.

5. Observe the arc and the electrodes to find out how great an area at each pole becomes incandescent (gives off bright light).

6. After thirty to sixty seconds of arc action, break the arc and observe both electrodes closely to see what effect the arc action has had on both the negative and positive carbons.

Observations

1. At which pole of the DC arc was the greatest amount of heat released?

2. About how much heat was released at each pole? On what is this conclusion based?

3. What was the condition of the electrodes after completing the experiment?

4. If the poles had been reversed what would the result be?

EXPERIMENT 2

Materials

DC welding machine
Sharpened carbon electrode 1/4-inch diameter
Mild steel plate about 4 inches square and 1/4 inch to 1/2 inch thick

Procedure

1. Set the machine to deliver 150 or more amps., DCSP.

2. Draw an arc and melt the metal to form a bead about 1 1/2 inches long. Observe the arc action during the welding and inspect the electrode after welding.

3. Reverse the welding current using DCRP.

4. Run a second bead some distance from the first and parallel to it, observing the arc action and the condition of the electrode after completing this weld.

5. Center punch the following using a hammer and center punch.

   a. the plate
   b. the weld made with DCSP
   c. the weld made with DCRP

   CAUTION: Be sure to wrap the pointed end of the punch with cloth, a leather glove or other material that will guard against flying metal should the punch break. Because of the hazard involved, the test described in step 5 should be conducted by the instructor.

Observations

1. What was observed about the ease or difficulty of welding in this experiment?

2. How did the finished beads compare?

3. What was the condition around the welds in regard to residue?

4. What observations and conclusions were made when the plate and the DCSP weld were center punched?

5. When the DCRP weld was center punched, what conclusions were made?

6. Is resistance to penetration a good test of hardness? Explain.

7. From observations made in this experiment, is the greatest mass action in the carbon arc from negative to positive or from positive to negative? Explain.

EXPERIMENT 3

Materials

DC welding machine
Sharpened 1/4-inch diameter carbon electrode
Aluminum plate about 2 in. x 6 in. x 1/8 in.

Procedure

1. Strike an arc on the plate and run a bead using about 150 amps., DCSP. Observe the arc action and, in particular, the molten aluminum. A wrinkled, dull-appearing skin will form over the molten pool.

2. Examine the finished bead for appearance and note the condition of the electrode.

3. Repeat step 1 but use DCRP. Note the size of the pool of molten metal directly under the arc as compared with that found with DCSP. Also compare the brightness of this pool.

Observations

1. What was observed about the amount of penetration when using DCSP in this experiment? Is this consistent with the findings in Experiment 1?

2. What surface condition was observed directly under the arc when using DCRP?

3. What conclusion can be drawn from the above?

4. What accounts for the black residue on the aluminum when using DCRP, as this condition was not observed when welding the steel with the same polarity?

5. What was observed about the amount of penetration and the size of the bead in this experiment? What significance is attached to this observation?

THE ALTERNATING CURRENT ARC

In the direct-current welding circuit the flow of current across the arc is always from (−) to (+). In the alternating current arc, the current reverses itself many times a second, which means that alternating current is a combination of DCSP and DCRP. The most common welding frequency is 60-cycle AC. In 60-cycle AC, the electrode is positive (+) 60 times each second and negative (−) 60 times each second. It is also momentarily zero 120 times each second, a fact which makes the AC arc difficult to maintain. In this case, a high-frequency AC carrier or igniter arc is superimposed on the standard 60-cycle AC welding current to produce HFAC which eliminates the problem of maintaining an arc.

EXPERIMENT 4

Materials

AC welding machine
Sharpened 1/4-inch diameter carbon electrode
Aluminum plate about 2 in. x 6 in. x 1/8 in.

Procedure

1. Strike the arc, using about 150 amps., and again observe the arc. Note the size, shape and degree of brightness of the molten metal.

2. Try varying the arc length, making continuous observations. Note the ease or difficulty of maintaining the arc.

Observation

1. What conclusions can be drawn from this experiment?

REVIEW QUESTIONS

This unit has led the students through a series of experiments and questions designed to help them form conclusions that are important to the study of TIG welding. What can be concluded regarding:

1. Oxidation of aluminum?

2. Effect of straight polarity?

3. Effect of reverse polarity?

4. Comparison of AC with DC with regard to penetration and cleaning?

5. Carbon transfer?

FUNDAMENTALS OF TIG WELDING

Although it can produce outstanding results, the TIG welding process may be unnecessarily expensive. A careless operator can cause a major expense by damaging the equipment. In particular, the tungsten electrode and the ceramic nozzle are subject to misadjustment. Care must be taken with these parts, therefore, since they are important to the production of high-quality work at a moderate cost.

COMPARISON OF METAL ARC WELDING WITH TIG WELDING

The variables found in TIG welding are almost identical to those in metal arc welding. These variables are

- Length of arc
- Amount of current
- Rate of arc travel
- Angle of electrodes

The student who is skilled in the metal arc welding process is already familiar with these variables. This prior knowledge is helpful in the study of TIG welding.

The following differences in costs must be considered:

- The overall value of the equipment used in TIG welding is much higher. This includes the welding machine, the cable and hoses, as well as the torch, regulator and nozzles.

- The shielding gas used is much more expensive than gases used in most other welding processes. For example, helium and argon gases are far more expensive than acetylene.

- The electrodes used in TIG welding are much more expensive. Actually, these electrodes are consumed so slowly that the cost of electrodes per foot of weld is very slight. However, the student should realize that any waste of electrodes due to bending, breaking or the use of excessive current is a very expensive error.

- The material being welded is generally much more expensive and sometimes hard to obtain, especially in the sizes and alloys desired.

In general, the student of TIG welding should be aware of the costs involved and should follow an intelligent, rigid procedure to protect the equipment from costly damage and to avoid the waste of expensive materials.

ELECTRODES

Tungsten and tungsten alloys are supplied in diameters of .010 inch, .020 inch, .040 inch, 1/16 inch, 3/32 inch, 1/8 inch, 5/32 inch, 3/16 inch, and 1/4 inch. They are manufactured in lengths of 3 inches, 6 inches, 7 inches, 18 inches, and, in some instances, 24 inches. The electrodes are made with a cleaned surface, either chemically cleaned and etched, or with a ground finish which holds the diameter to a closer tolerance. Electrodes are supplied in pure tungsten and in three alloys: 1 percent thorium, 2 percent thorium alloy and zirconium alloy.

Chart 6-1

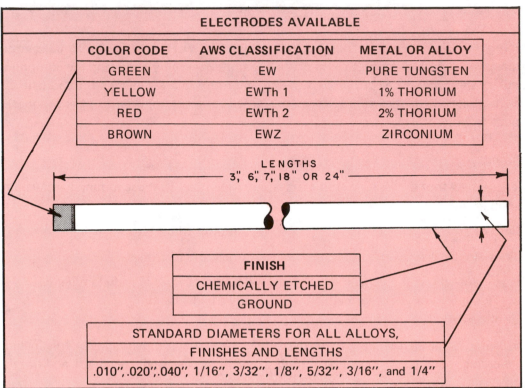

ELECTRODES AVAILABLE		
COLOR CODE	AWS CLASSIFICATION	METAL OR ALLOY
GREEN	EW	PURE TUNGSTEN
YELLOW	EWTh 1	1% THORIUM
RED	EWTh 2	2% THORIUM
BROWN	EWZ	ZIRCONIUM

LENGTHS
3", 6", 7", 18" OR 24"

FINISH
CHEMICALLY ETCHED
GROUND

STANDARD DIAMETERS FOR ALL ALLOYS,
FINISHES AND LENGTHS
.010", .020", .040", 1/16", 3/32", 1/8", 5/32", 3/16", and 1/4"

Pure tungsten is generally used with AC welding. The thoriated types are mostly used for DCSP welding and give slightly better penetration and arc starting characteristics over a wider range of current values. The zirconium alloy is excellent for AC welding and has high resistance to contamination. Its chief advantage is that it can be used in those instances when contamination of the weld by even very small quantities of the electrode is absolutely intolerable.

Chart 6-1 condenses information on electrodes available for TIG welding. The color code shown is being used by the major producers and distributors of tungsten electrodes.

The recommended amperage for any given size electrode varies with the type of joint being welded and the type of current used. A general recommendation when welding with AC is that the current be equal to the diameter of the electrode in thousandths of an inch multiplied by 1.25. For example, an electrode with a diameter of .040 requires a current of 40 x 1.25 or 50 amperes. Of course, the size of the electrode is a function of the thickness of the metal being welded. Chart 6-2 gives current ratings and electrode sizes for butt welding the various thicknesses of aluminum using HFAC arc with argon shielding and pure tungsten electrodes.

While charts are valuable as guides, a degree of sound judgment on the part of the operator is also desirable. Electrodes operated at a current value which is too low cause an erratic arc just as with metal arc welding. If the current is correct, the end of the electrode appears as in figure 6-1.

Chart 6-2

DATA FOR STRINGER BEADS IN ALUMINUM						
Thickness in Inches	HFAC Welding Current Flat* Amperes	Tungsten Electrode Diameter	Welding Speed Inches per Min.	Filler Rod Diameter	Recommended Argon Flow Cu. Ft. per Hr. ***	Gas Nozzle Size
1/16	60-80	1/16	12	1/16	15 to 20	4, 5, 6
1/8	125-145	3/32	12	3/32 or 1/8	17 to 25	6, 7
3/16	190-220	1/8	11	1/8	21 to 30	7, 8
1/4	260-300**	3/16	10	1/8 or 3/16	25 to 35	8, 10
3/8	330-380**	3/16, 1/4	5	3/16 or 1/4	29 to 40	10
1/2	400-450**	3/16, 1/4	3	3/16 or 1/4	31 to 40	10

* — Current values are for flat position only. Reduce the above figures by 10% — 20% for vertical and overhead welds.

** — For current values over 250 amps., use a torch with a water-cooled nozzle.

*** — Use lower argon flow for flat welds. Use higher argon flow for overhead welds.

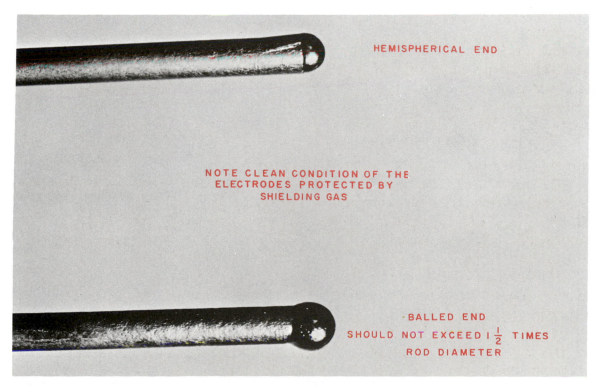

HEMISPHERICAL END

NOTE CLEAN CONDITION OF THE ELECTRODES PROTECTED BY SHIELDING GAS

BALLED END SHOULD NOT EXCEED 1½ TIMES ROD DIAMETER

Fig. 6-1 Electrodes operated at proper current

Note that this figure shows a round, shiny end in one case and an end which forms a ball in the other. If this ball is over one and one-half times the diameter of the electrode, the current is too high and the electrode is consumed at an excessively high rate.

CERAMIC NOZZLES

The ceramic nozzles in chart 6-2 are indicated in fractions of an inch to describe the recommended inside diameter of these nozzles. There is a trend in the industry to indicate the nozzles by numbers such as 4, 5, 6, and 7. These numbers give the nozzle size in sixteenths of an inch. For example, a #6 nozzle indicates 6/16-inch or 3/8-inch inside diameter.

In general, the inside diameter or orifice of the nozzle should be from four to six times the diameter of the electrode. Nozzles with an orifice which is too small tend to overheat and either break or deteriorate rapidly. Smaller-diameter nozzles are also more subject to contamination. Ceramic nozzles are usually recommended for currents up to 250-275 amps. Above this point, special torches with water-cooled metal nozzles are generally used.

One other important factor is the amount the electrode extends beyond the nozzle, figure 6-2.

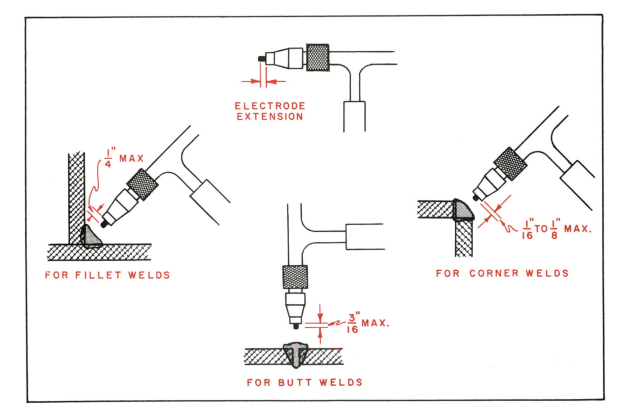

Fig. 6-2 Electrode extension

REVIEW QUESTIONS

1. Zirconium electrodes cost fifty percent more than tungsten but have excellent characteristics for HFAC welding. Would it be justifiable to use them?

2. Chart 6-2 shows the shielding gas flow for various sizes of electrodes. When welding aluminum would it be justifiable to experiment with this factor? On what basis?

3. Why would it be desirable to use seven-inch tungsten electrodes whenever possible instead of three-inch electrodes?

4. When would it be justifiable to use three-inch tungsten electrodes?

5. What is the objection to a long electrode extension?

6. What current in amperes is required for a 1/16-inch electrode?

Unit 7

STARTING AN ARC AND RUNNING STRINGER BEADS ON ALUMINUM

Rigid attention to detail and procedure is of extreme importance in TIG welding. Errors due to carelessness may prove to be very expensive. For instance, failure to turn on the cooling water usually results in destruction of the torch as well as the cable and hose assembly. Striking an arc with the machine set for normal amperage but with the polarity selector on DCRP will result in destruction of the electrode and usually the collet holder and collet.

Materials

Clean aluminum plate 1/8 in. thick x 4 in. x 6 in.
AC welding machine equipped with high-frequency oscillator
Cylinder of argon gas equipped with flowmeter
TIG torch fitted with 1/8-inch pure tungsten electrode and a #7 or #8 nozzle, (7/16 inch or 1/2 inch.)

Preweld Procedure

1. Make sure the torch is well away from the ground or work cable.

2. Turn on the cooling water.

3. Set the high-frequency switch to AC TIG.

4. Set water and gas switch to ON position.

5. If remote control is used, set switch to ON or remote. Otherwise leave it at LOCAL or OFF.

6. If the machine is equipped with a balanced wave filter or batteries, set this switch to ON.

7. Set gas and water afterflow timer for 1/8-inch electrodes.

8. Turn on the gas from the argon cylinder and adjust the flowmeter to supply 17 to 21 cubic feet per hour. (Read to the top of the ball.)

   Note: The flow of argon and the flow of water cannot be checked unless the remote control switch is ON or, if local control is used, the machine power switch is in the ON position.

9. Set the polarity switch to AC.

10. Adjust the current as indicated in chart 6-2.

11. Check the electrode for the proper extension. Refer to figure 6-2.

12. With the power OFF, check the electrode to be sure it is firmly held in the collet. To do this, place the exposed end of the electrode against a solid surface and push the torch down gently but firmly. If the electrode tends to move back into the nozzle, either the collet holder or gas cap needs to be tightened. Be sure to set the electrode extension and then tighten the collet holder or gas cap.

13. Turn the power switch ON. Turn the remote switch to ON. Note that the flowmeter indicates the proper flow of shielding gas. Check the waste line to be sure the cooling water is flowing.

14. Strike an arc by bringing the electrode close to a grounded workpiece, preferably copper. If the arc fails to jump from the electrode to the work without actual contact, set the high-frequency intensity control to a higher setting.

15. Again strike an arc on the copper and allow the current to flow until the electrode becomes incandescent. Break the arc and check the afterflow by watching the electrode cool. The instant the electrode becomes bright, look at the small ball in the flowmeter tube. For example, a 1/8-inch electrode has a diameter of 0.125 inch. Therefore, the postflow is 12.5 seconds for 1/8-inch electrodes.

Note: If the afterflow timing was originally set for too short duration, the electrode would cool in the atmosphere and oxidize. This is indicated by the electrode becoming blue or black in color. In this event, adjust the afterflow timer higher until the ideal afterflow is reached.

The above procedure should be strictly followed each time the TIG welding process is used, even if someone else has just completed a weld with the equipment. It is the personal responsibility of each operator to be sure that all controls are properly adjusted at all times. Carelessness can lead to serious damage to the equipment.

Welding Procedure

1. Use all standard safety precautions. In addition, use a welding helmet filter plate one or two shades darker than the one for metal arc welding (#11 or #12).

2. Place the clean aluminum workpiece in firm contact with a clean worktable surface. The radio-frequency high-voltage (2000-4000 volts) igniter arc will jump a wide gap. If it jumps from the worktable to the reverse side of the workpiece, it may cause damage to the surface finish.

3. There are times when backing bars are advantageous. Either steel or stainless steel is recommended for backing bars; stainless steel is the better choice. Copper is not good for a backing bar when welding aluminum, and carbon should never be used for this purpose.

4. Turn on the remote switch and draw an arc with the electrode held as nearly vertical as possible while observing the molten puddle, figure 7-1. Use an arc about equal in length to the electrode diameter.

5. Run a straight bead about 1/2 inch from, and parallel to, the edge of the plate, with the rate of the arc travel adjusted to maintain a pool of molten metal about 3/8 inch in diameter.

6. Examine the finished bead for uniformity and surface appearance. The weld should have a shiny appearance along its entire length. Note that there is an area about 1/8 inch on each side of the weld which is quite white but dull in lustre, figure 7-2. This

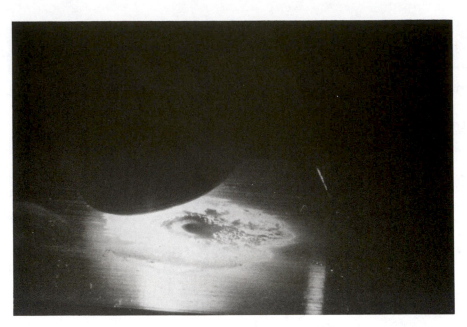

Fig. 7-1 Starting an arc

Fig. 7-2 Partially vaporized oxide, no preweld cleaning

is aluminum oxide which has been partially vaporized by the high-frequency igniter arc. Also examine the electrode for brightness and shape of the arc. Examine the reverse side of the plate for penetration.

7. Run a second bead 1/2 inch from the first and parallel to it. While making this bead, pay close attention to the area just outside the molten pool. A great number of pinpoint-size arcs should be seen. This is the high-frequency arc partially breaking up the aluminum-oxide film.

8. Examine the finished bead for surface appearance, uniformity, penetration and for the amount of white residue along the edges of the weld. Check the bead for cracking.

9. Make a third and fourth bead, observing the arc action and adjusting the rate of arc travel to obtain a uniform weld with complete penetration.

10. Examine these beads critically as in step 8. Examine the electrode after each bead for evidence of excessive burn-off or electrode contamination.

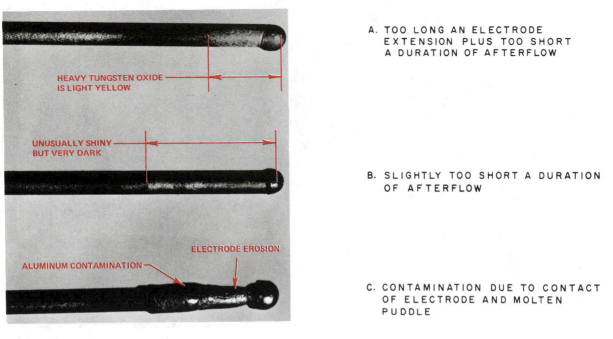

A. TOO LONG AN ELECTRODE EXTENSION PLUS TOO SHORT A DURATION OF AFTERFLOW

B. SLIGHTLY TOO SHORT A DURATION OF AFTERFLOW

C. CONTAMINATION DUE TO CONTACT OF ELECTRODE AND MOLTEN PUDDLE

HEAVY TUNGSTEN OXIDE IS LIGHT YELLOW

UNUSUALLY SHINY BUT VERY DARK

ELECTRODE EROSION

ALUMINUM CONTAMINATION

Fig. 7-3 Electrode contamination

Figure 7-3, view A, shows an electrode contaminated from the atmosphere because of too short a duration of afterflow of the shielding gas plus excessive electrode extension. In view B, an electrode is shown that was contaminated because of too short a duration of afterflow of shielding gas. View C shows an electrode contaminated by allowing it to come in contact with the molten pool.

11. Make a fifth bead but this time allow the electrode to contact the molten pool two or three times as the weld progresses. Observe the arc action and the area around the weld, especially after contact has been made.

12. Observe the finished bead as in step 7 and note the difference. Examine the electrode and gas nozzle for evidence of contamination as in figure 7-3, view C.

    Note: In the interest of electrode economy, the aluminum contamination of the electrode can be burned off by allowing the arc to dwell for several seconds on a copper plate. This is acceptable for practice welding. However, for high-production, high-quality welds, the accepted practice is to remove the electrode and either grind away the contaminated area or notch the electrode just back of the contamination and break it off. If this is done, be sure to grasp the electrode close to the notch to avoid bending.

13. Do the experiment shown in figure 7-4. After the aluminum has broken cool the pieces pieces and examine them. Observe that there is no evidence of melting. This indicates that the metal has broken rather than melted. This phenomenon, termed *hot-short,* is not uncommon. Many metals and alloys, such as copper, brass, and even cast iron display this characteristic at a temperature slightly below their melting point. Consider what

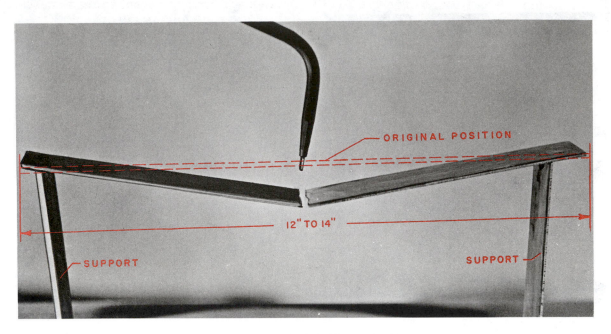

ORIGINAL POSITION

12" TO 14"

SUPPORT

SUPPORT

Fig. 7-4 Aluminum breaking under heat

would happen if the aluminum strip were laid on a flat steel plate and heated. Would the piece have broken or simply melted? What conclusion can be drawn from the above?

REVIEW QUESTIONS

1. What four steps in the preweld procedure, in order of their importance, are most essential in avoiding damage to the equipment or electrodes?

2. The inert gas used with the TIG process is registered on a flowmeter according to what unit of measurement?

3. What is the function of the postflow timer?

4. What two metals are recommended for use as backing bars? Of the two metals, which is better?

5. Compare the amount of the white or light residue observed adjacent to the first weld with that observed in the succeeding welds. How do the welds differ?

6. What conclusion can be drawn from the observation in Question 5?

7. What practical use is there for the phenomenon observed in Questions 5 and 6?

8. What observations were made when welding with an electrode contaminated with aluminum?

9. What conclusions can be drawn from the answer to Question 8?

10. What conclusions can be drawn from the experiment in step 13?

Unit 8

RUNNING PARALLEL STRINGER BEADS ON ALUMINUM

This unit provides the first opportunity to use TIG welding in which additional metal is added to the bead by using a filler rod. Those who have welded with a carbon arc using a filler rod will find much similarity between TIG welding and the processes and techniques with which they are familiar.

Materials

Aluminum plate 1/8 in. x 4 in. x 6 in. to 9 in. long, free from oil, grease and lint
1/8-inch aluminum or 5% silicon-alloy rod

Procedure

1. Choose the proper size tungsten electrode, nozzle and gas flow from chart 6-2.
2. Follow the preweld procedure in unit 7.
3. Make a second check of the four most important factors.

   a. Water flow
   b. Polarity of electrode
   c. Flow of shielding gas
   d. Firmness of contact between the electrode and collet

4. Strike an arc and run a bead about 3/8 inch wide parallel to the edge of the aluminum plate and about 1/2 inch from the edge, holding the torch and filler rod as shown in figure 8-1.
5. Examine the finished bead for fusion, uniformity and completeness of penetration, and white residue.

   Note: Since there is no requirement that either the plate or filler rod be cleaned prior to welding, some difficulty may have been found in fusing the filler rod and plate. This is due to the tough skin or film of aluminum oxide on the work.

   Oxyacetylene welders of aluminum overcome this difficulty by alternately jabbing the filler rod into the molten pool and then withdrawing it in a regular, rhythmic motion. This allows the rod to break through the oxide film at each down cycle. TIG welding uses a different procedure whereby the filler rod is brought to the leading edge of the molten puddle and the torch is moved steadily in a forward direction. Care has to be taken that the tungsten does not make contact with the rod or molten puddle. This would result in electrode contamination and the difficulties described in unit 7.

6. Run beads until the technique shown in figure 8-2 is perfected.
7. Examine each bead for appearance, penetration and residue, figure 8-3.
8. Check the electrode after each bead for electrode extension and contamination.

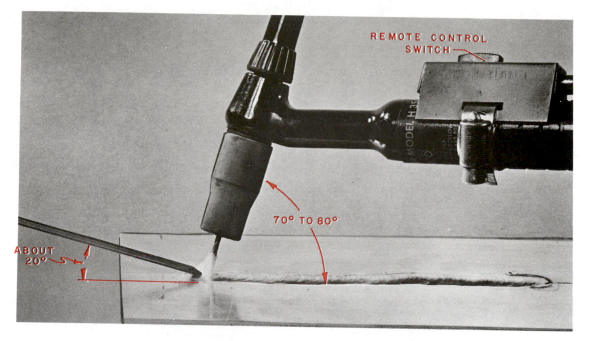

Fig. 8-1 Relative position of torch and filler rod

Fig. 8-2 Move the arc back as the filler rod is added.

Fig. 8-3 Parallel stringer beads on aluminum

9. Obtain more aluminum plates and clean them prior to welding, either with steel wool or abrasive cloth.

10. Run a bead on this cleaned plate as in steps 4 and 6. Pay close attention to the ease or difficulty of fusing the filler rod and work.

11. Examine the finished bead as before. Be sure to compare the amount of white residue along the edges of this weld with that observed in steps 5 and 7.

12. Cool the plate and again clean it in the area where the next bead is to be made. Also clean the filler rod thoroughly and run another bead. Note the ease or difficulty of fusing the filler rod and workpiece.

13. Examine this bead, paying close attention to the amount of white residue along the edges of the weld as compared to that along the edges of beads observed in steps 5,7, and 11.

14. Make more beads with both the plate and filler rod. Cool and clean the plate and rod before each bead. Try varying the angle of the filler rod. Remember that the height of the bead is a function of the angle between the filler rod and the work as in oxyacetylene welding. Also try varying the angle of the work and electrode slightly. If the torch angle is too low the argon gas draws some air along with it and breaks down the shielding qualities, figure 8-4.

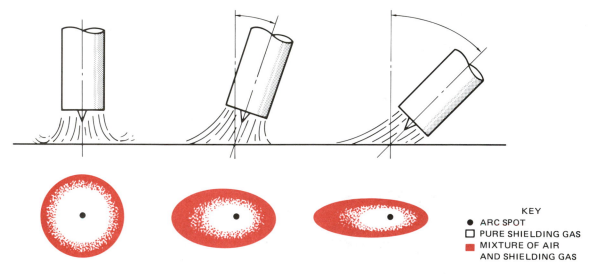

KEY
● ARC SPOT
☐ PURE SHIELDING GAS
■ MIXTURE OF AIR AND SHIELDING GAS

Fig. 8-4 Gas coverage patterns for different GTA torch angles. Note how the area covered by the shielding gas becomes narrower and elongates as the angle of the torch increases from the perpendicular. (Reprinted from Jeffus & Johnson, *Welding: Principles & Applications,* Figure 15-1)

REVIEW QUESTIONS

1. What are the proper filler rod and torch angles for TIG welding?

2. When using chart 6-2 to find the amperage setting and flow of shielding gas, would the higher current setting and gas flow be used or the lower? Why?

3. What are the objections to using aluminum plate which has not been cleaned of oil, grease, and dust?

4. What harmful effects does failure to clean the oxide from the work and filler rod have on the finished weld?

5. In observing the reverse side of the welded plates, it is found that the surface has a dull, wrinkled appearance. How can this be overcome?

6. Is abrasive cleaning of the work and filler rod worthwhile? Why?

7. When would abrasive cleaning be undesirable? What cleaning method could be used?

Unit 9

OUTSIDE CORNER WELDS ON ALUMINUM

This unit provides the opportunity to make high-quality welds of excellent appearance without the responsibility of manipulating a filler rod. It also gives the added opportunity to construct and use a simple jig or welding fixture. In all following units, material designated as aluminum is understood to mean either pure aluminum, if available, or any of its weldable alloys.

Outside corner welds on aluminum (see figure 9-3) require the construction of a jig, either of steel or stainless steel, as shown in figure 9-1. Be sure that all oxides and dirt are cleaned from this jig before using. A better jig can be made by adding clamps and clamping bars as shown in figure 9-2.

Materials

2 pieces 1/8 in. x 1 1/2 in. x 6 in. to 9 in. long aluminum or weldable aluminum alloy

Procedure

1. Place the two workpieces in the jig in such a way that their edges are brought in close contact along their entire length as shown in figure 9-4.

2. Before welding make a complete check on all welding factors as in unit 7. Be sure to refer to figure 6-2 to find the proper electrode extension.

4"x 4"x $\frac{1}{2}$" ANGLE IRON
BOTH LEGS MACHINED
SQUARE

9"

THIS EDGE ROUNDED
SLIGHTLY

Fig. 9-1 Simple jig

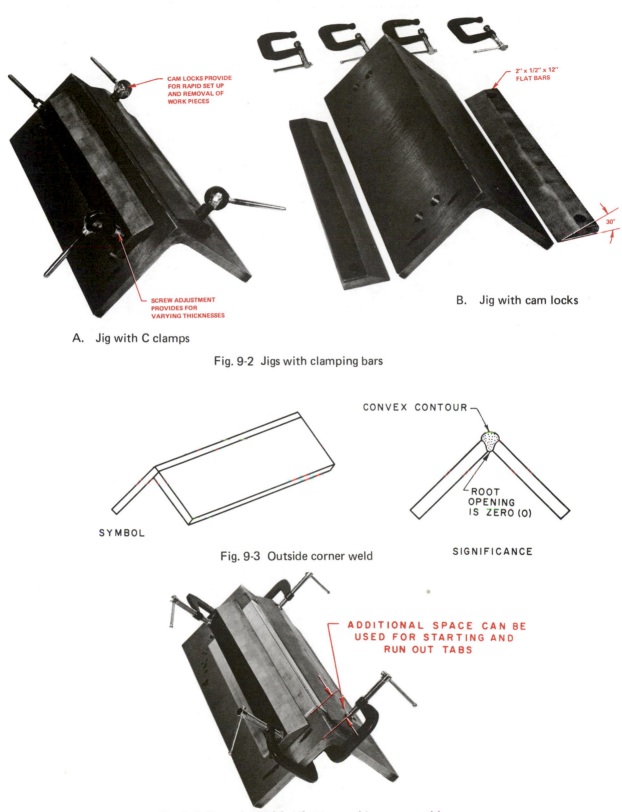

CAM LOCKS PROVIDE
FOR RAPID SET UP
AND REMOVAL OF
WORK PIECES

SCREW ADJUSTMENT
PROVIDES FOR
VARYING THICKNESSES

2" x 1/2" x 12"
FLAT BARS

30°

A. Jig with C clamps

B. Jig with cam locks

Fig. 9-2 Jigs with clamping bars

SYMBOL

CONVEX CONTOUR

ROOT
OPENING
IS ZERO (0)

SIGNIFICANCE

Fig. 9-3 Outside corner weld

ADDITIONAL SPACE CAN BE
USED FOR STARTING AND
RUN OUT TABS

Fig. 9-4 Plates in position for an outside corner weld

Chart 9-1

DATA FOR CORNER WELDS IN ALUMINUM					
Thickness in Inches	HFAC Welding Current Flat* Amperes	Tungsten Electrode Diameter	Welding Speed Inches per Min.	Filler Rod Diameter	Recommended Argon Flow Cu. Ft. per Hr. ***
1/16	60-80	1/16	12	1/16	15 to 20
1/8	125-145	3/32	12	3/32	17 to 25
3/16	190-220	1/8	11	1/8	21 to 30
1/4	280-320**	3/16	10	1/8 or 3/16	25 to 35
3/8	350-400**	3/16, 1/4	5	3/16 or 1/4	29 to 40
1/2	420-470**	3/16, 1/4	3	3/16 or 1/4	31 to 40

* — Current values are for flat position only. Reduce the above figures by 10% — 20% for vertical and overhead welds.

** — For current values over 250 amps., use a torch with a water-cooled nozzle.

*** — Use lower argon flow for flat welds. Use higher argon flow for overhead welds.

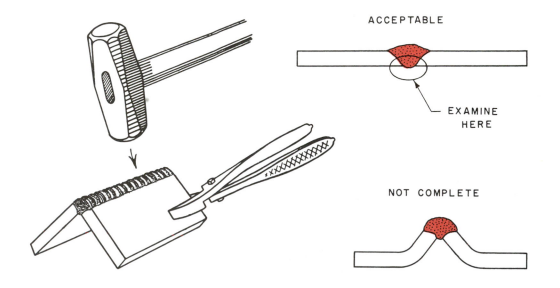

Fig. 9-5 Testing an outside corner weld

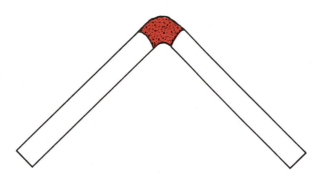

Fig. 9-6 Root of weld with slight fillet

3. Refer to chart 9-1 for electrode size, amperage, and argon flow.

4. Strike an arc and weld the corner; try for a smooth joint with **100** percent fusion.

5. Cool and examine the weld and the reverse side of the joint for uniformity and fusion. Also note any residue.

6. To test this weld place it on an anvil and hammer it flat as shown in figure 9-5.

7. Clean two more plates in the manner used in unit 8 (with steel wool or abrasive cloth) and place them in the jig.

8. Make this weld and be sure to note any change in the ease or difficulty of welding as compared to step 4.

9. Cool and inspect this weld for appearance, fusion, residue and evidence of wrinkled aluminum oxide on the root side of the assembly. Compare with the joint previously made without cleaning the material.

10. Make more corner welds and test them as in step 6. If it is found that there is a tendency for the joint to break, remove a little material from the corner of the jig by filing or grinding so that the root of the weld can take the shape indicated in figure 9-6.

11. If polishing and buffing equipment are available, try buffing a joint and examine it to find out how a rough weld, or a weld with deep ripples, tends to increase the buffing time and cost of finishing.

REVIEW QUESTIONS

1. What care should be taken of the jig once it is put in use? Why?

2. What condition leaves the root side of the joint in the best appearance?

3. What effect does a slight fillet have on the ability of the joint to withstand hammer testing?

4. If it is necessary to provide a joint with the root side absolutely free from oxide, how would this be accomplished?

5. What is the best material to use when making the jig? Why?

BUTT WELDS ON ALUMINUM

This unit provides the opportunity to gain skill and knowledge in making what is probably the most common joint used in welding, the butt joint. The butt weld is shown in figure 10-1.

Materials

2 pieces 1/8 in. x 2 in. x 6 in. to 9 in. long clean aluminum
1/16 in. or 3/32 in. x 36 in. long E4043 welding rod

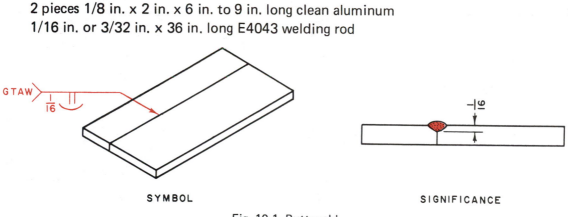

SYMBOL   SIGNIFICANCE

Fig. 10-1 Butt weld

Procedure

1. To get a better job, use a backing bar as shown in figure 10-2, rather than tacking and welding the aluminum on a welding bench. A steel bar is acceptable but stainless steel is better.

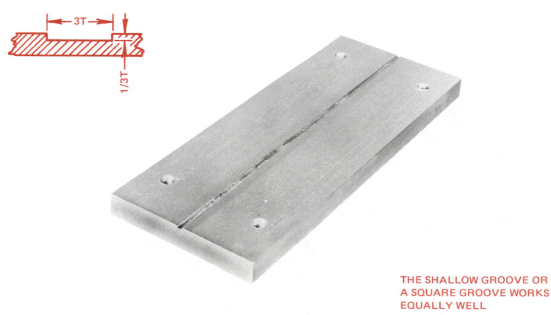

THE SHALLOW GROOVE OR
A SQUARE GROOVE WORKS
EQUALLY WELL

Fig. 10-2 Backing bar for welding aluminum

Chart 10-1

DATA FOR BUTT WELDS IN ALUMINUM					
Thickness In Inches	HFAC Welding Current Flat* Amperes	Tungsten Electrode Diameter	Welding Speed Inches per Min.	Filler Rod Diameter	Recommended Argon Flow Cu. Ft. per Hr. ***
1/16	60-80	1/16	12	1/16	15 to 20
1/8	125-145	3/32	12	3/32 or 1/8	17 to 25
3/16	190-220	1/8	11	1/8	21 to 30
1/4	260-300**	3/16	10	1/8 or 3/16	25 to 35
3/8	330-380**	3/16, 1/4	5	3/16 or 1/4	29 to 40
1/2	400-450**	3/16, 1/4	3	3/16 or 1/4	31 to 40

* — Current values are for flat position only. Reduce the above figures by 10% — 20% for vertical and overhead welds.

** — For current values over 250 amps., use a torch with a water-cooled nozzle.

*** — Use lower argon flow for flat welds. Use higher argon flow for overhead welds.

2. Use the standard preweld procedure before starting to weld.

3. Refer to chart 10-1 for electrode size, current setting, and argon gas flow.

4. Check the electrode extension and the seating of the electrode in the collet.

5. Make a bead, figure 10-3, using procedure and technique similar to those described for making stringer beads in unit 8.

Fig. 10-3 Filler rod being melted before it is added to molten puddle (Reprinted from Jeffus & Johnson, *Welding: Principles & Applications,* Figure 15-14A)

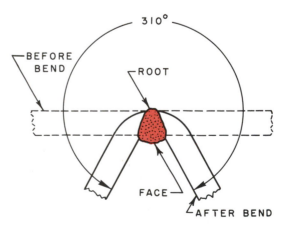

Fig. 10-4 Bend test for butt weld

6. Cool and examine the finished bead for appearance and complete fusion.

7. Test the finished weld by bending it in a brake or vise as shown in figure 10-4.

8. Examine the root side of the bent piece. There should be no indication of cracking or failure.

9. Continue to make tests and examine butt welds until the welds made are consistently of excellent quality and appearance with a cross section similar to that shown in figure 10-5.

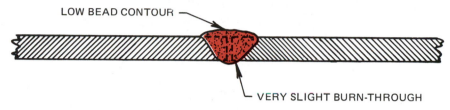

LOW BEAD CONTOUR

VERY SLIGHT BURN-THROUGH

Fig. 10-5 Cross section of ideal butt weld

10. It is not always practical to make butt welds using a backing bar. Try tacking some plates and supporting them so that there is an air space between the workbench and the root of the joint. Make butt welds using this procedure until the welded joints are of uniform appearance with good fusion at the root, but not too great an amount of burn-through.

    Note: If there is some difficulty in fusing this type of joint, try the technique shown in figure 8-2.

11. Test the fusion at the root of the weld by bending these welds as in step 7.

REVIEW QUESTIONS

1. What dimensions should the groove in a backing bar have?

2. What will the result be at step 10 if the filler rod is allowed to move far enough to be outside the shielding gas zone?

3. What are the recommendations for joint spacing for a square butt joint in 1/8-inch plate?

4. If the surface of this bead is white and dull what does it indicate?

5. What is the thickest material that can obtain 100 percent penetration from one side?

LAP WELDS ON ALUMINUM

While lap welds, figure 11-1, and fillet welds (which will be studied in unit 12) have many similarities, there are enough differences to make the separate study of each joint worthwhile.

Materials

2 pieces clean aluminum, 1/8 in. x 2 in. x 6 in. to 9 in. long
1/16 in. or 3/32 in. x 36 in. long E4043 welding rod

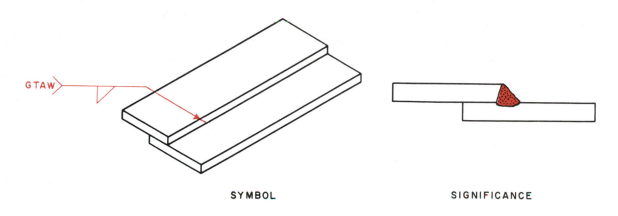

SYMBOL   SIGNIFICANCE

Fig. 11-1 Lap weld

Procedure

1. Follow the standard preweld procedure for TIG welding equipment.

2. Set up the cleaned plates as shown in figure 11-2.

3. Refer to chart 11-1 for electrode size, amperage setting, argon gas flow and nozzle size.

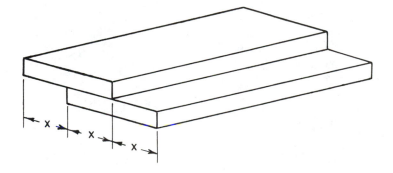

Fig. 11-2 Setup for lap joint

47

Chart 11-1

	DATA FOR LAP AND FILLET WELDS IN ALUMINUM				
Thickness in Inches	HFAC Welding Current Flat* Amperes	Tungsten Electrode Diameter	Welding Speed Inches per Min.	Filler Rod Diameter	Recommended Argon Flow Cu. Ft. per Hr. ***
1/16	70-90	1/16	10	1/16	15 to 20
1/8	140-160	3/32	10	1/16 or 3/32	17 to 25
3/16	210-240	1/8	9	1/8	21 to 30
1/4	280-320**	3/16	8	1/8 or 3/16	25 to 35
3/8	330-380**	3/16, 1/4	5	3/16 or 1/4	29 to 40
1/2	400-450**	3/16, 1/4	3	3/16 or 1/4	31 to 40

* — Current values are for flat position only. Reduce the above figures by 10% — 20% for vertical and overhead welds.

** — For current values over 250 amps., use a torch with a water-cooled nozzle.

*** — Use lower argon flow for flat welds. Use higher argon flow for overhead welds.

Note: Use an electrode extension which is just slightly longer than for butt welds.

4. Make a lap weld by forming a puddle on the bottom plate. Shorten the arc and play the torch over the upper plate, applying filler rod along the edge of the top plate. Pay close attention to the molten pool. A notch effect is usually observable in the joint as in figure 11-3. Avoid melting the edge of the top plate more rapidly than the center of the bottom plate.

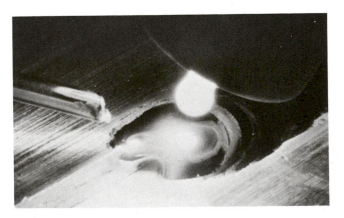

Fig. 11-3 Notch effect

5. Each edge of the weld tends to melt some distance ahead of the center of the pool, which results in a loss of the continuity of the weld. To overcome this, apply the technique used in making stringer beads and butt welds.

6. Make more lap welds, using both sides of the plates, except in those instances where it is desired to test the finished weld. Note that the electrode angle in figure 11-4 is much greater than that used for metal arc welding.

7. Test the finished welds as shown in figure 11-5.

8. Examine the root of the weld for evenness and uniformity of fusion.

9. Make more lap welds, but vary the technique to produce welds with a cross section similar to that in figure 11-6.

   Note: It has been proven by tests that a lap weld of this contour transfers stresses from one plate to the other more efficiently than a joint made with the face of the weld at a 45-degree angle.

10. Test these welds, figure 11-7, as in step 7, after first observing the finished bead for uniformity of fusion with both plates, as well as general appearance of the contour and ripples.

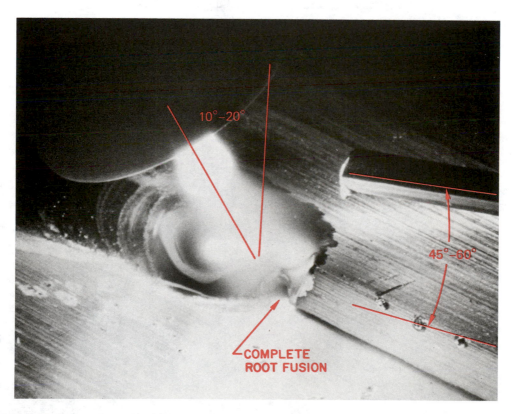

Fig. 11-4 Rod and torch angles (Adapted from Jeffus & Johnson, *Welding: Principles & Applications,* Figure 15-33A)

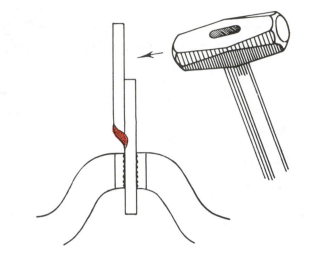
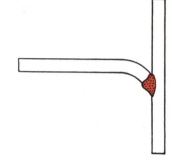

Fig. 11-5 Testing a lap weld

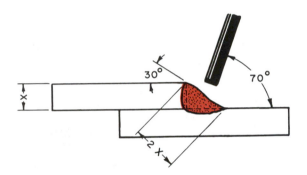

Fig. 11-6 Lap joint weld

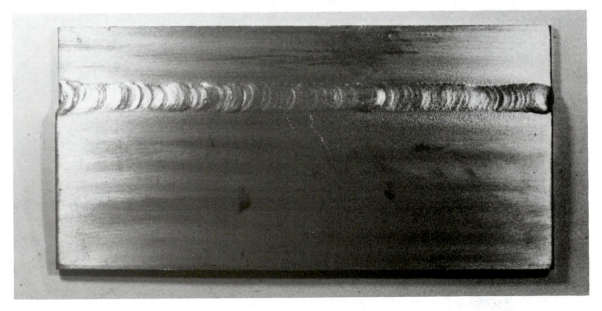

Fig. 11-7 Aluminum lap weld

REVIEW QUESTIONS

1. Should the current range for lap welds on aluminum be low, medium, or high? Why?

2. Where is the welding rod held in relation to the lap joint?

3. What rod angle affects melt-off rate?

4. What does the notch indicate in the weld crater?

5. Regarding penetration, what must be considered on this joint?

Unit 12

FILLET WELDS ON ALUMINUM

Fillet welds, figure 12-1, using the TIG torch may be slightly awkward to make. The gas nozzle gives some interference not found in either oxyacetylene or metal arc welding. However, attention to detail makes it possible to produce welds of excellent appearance and quality.

In making fillet welds care must be taken to avoid *undercutting,* figure 12-2, an absence of metal along the top edge of the weld. It is caused by faulty manipulation of the electrode and filler rod.

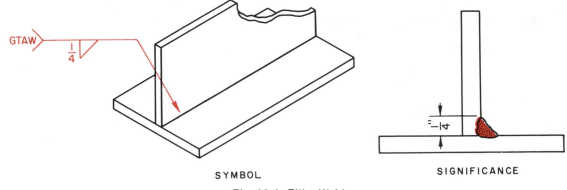

SYMBOL   SIGNIFICANCE

Fig. 12-1 Fillet Weld

Materials

2 pieces clean aluminum 1/8 in. x 2 in. x
6 in. to 9 in. long
1/16 in. or 3/32 in. x 36 in. long E4043
aluminum welding rod

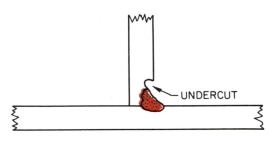

Fig. 12-2 Undercutting

Procedure

1. Check all factors of the equipment.

2. Set up the plates as in figure 12-1.

3. Refer to chart 11-1 for electrode size, nozzle size, amperage, and argon flow.

4. Refer to figure 6-2 and adjust the electrode extension.

5. Make a fillet weld with the torch angle and rod angle as shown in figure 12-3, view A.

6. Operate the electrode and filler rod as shown in figure 12-3, view B. Proper manipulation avoids the condition of undercutting which can happen as easily in TIG welding as in other processes.

7. Observe the finished weld for surface appearance and uniformity.

8. Test the weld as shown in figure 12-4. If the penetration and fusion are not deep enough, this joint will probably break. Examine the break for uniformity of the line of fusion.

   Note: Poor penetration is caused by not forming the weld pool in the base metal before adding filler rod to the leading edge of the weld pool.

(A) Welder's view

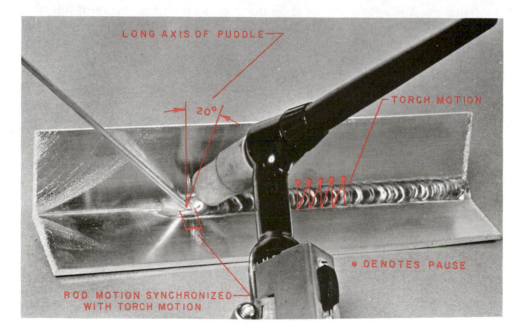

(B) Torch and rod motion

Fig. 12-3 Fillet weld (Adapted from Jeffus & Johnson, *Welding: Principles & Applications,* Figure 15-38)

Fig. 12-4 Testing a fillet weld

9. Make more fillet welds. Try to get deeper fusion and better bead appearance.

10. Continue to make and test this type of joint until the result is a high-quality weld every time.

REVIEW QUESTIONS

1. Why is undercutting considered such a serious error in the welding industry?

2. Is it possible to use a lower argon gas flow when making a fillet weld?

3. Would the afterflow time be the same duration as for other types of joints? Shorter? Longer?

4. Could it be possible to use a smaller ceramic cup on this joint? Why?

5. What shape does the surface of the bead present?

FLANGED BUTT WELDS ON ALUMINUM

Light-gage aluminum, as well as other thin metals and alloys, requires the maintenance of a rather short arc at a comparatively high rate of travel. The addition of filler rod to the joint further complicates the problem. For this reason, the flanged butt joint, figure 13-1, is often used when welding the lighter gage materials. This relieves the welder of the responsibility of manipulating the small-diameter, very flexible filler rod necessary to make a weld of acceptable appearance. The flanged butt weld is shown in figure 13-2.

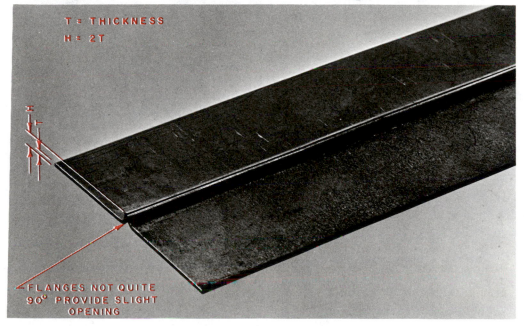

T = THICKNESS
H = 2T

FLANGES NOT QUITE 90° PROVIDE SLIGHT OPENING

Fig. 13-1 Flanged butt joint

2 pieces .030-inch to .050-inch clean aluminum, 2 in. wide x 9 in. long, flanged as shown in figure 13-1
Pure tungsten electrodes, .040 inch or 1/16 inch

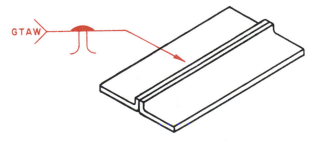

GTAW

SYMBOL

SIGNIFICANCE

Fig 13-2 Flanged butt weld

Fig. 13-3 Jig for welding light sheets

Procedure

1. Follow the standard preweld procedure for checking prior to HFAC TIG welding.

2. Use a .040-inch or 1/16-inch pure tungsten electrode with the machine set for the proper current value.

3. Set up the flanged sheets in a jig which holds the parts in proper alignment.

   Note: A steel or stainless steel jig such as that shown in figure 13-3 provides a quick and easy method of making the setup.

4. Strike the arc and make a weld, adjusting the rate of travel to provide good fusion and penetration.

5. Remove the finished work, cool, and examine both sides carefully for uniformity of the bead as well as for good fusion and uniform penetration.

6. Obtain more sheets and make more welds until each one made is of acceptable uniform appearance.

   Note: At this point, it is advisable to add starting and run-out tabs as shown in figure 13-4. Tabs are often used in high-speed welding to provide for a more uniform bead, with easier starting and stopping.

7. Set up more flanged plates, but change the electrode to one that is four times as large as previously used. Set the machine for DCRP electrode positive. Leave the current setting as it was for the small electrode used with the previous joints.

Fig. 13-4 Flanged butt joint with starting and run-out tabs

Note: If the torch is equipped to handle electrodes only 5/32 inch in diameter, do not exceed the current setting used for a .040-inch diameter electrode when using HFAC.

8. Make a weld using DCRP and note the ease or difficulty of producing the weld. Compare the rate of travel necessary with that required when using HFAC. Use a slightly longer arc when welding this joint.

9. Remove the work and examine it as in step 5.

10. Obtain some 1/16-inch clean aluminum plates and make square butt joints with DCRP.

11. Remove and examine these joints. Pay close attention to the width of the beads and the penetration at the root of the weld. Also, examine the electrode and compare its condition with the electrode used for HFAC welding.

12. Tack two flanged sheets of aluminum and weld them without the use of a jig. Compare the ease or difficulty of making this joint with the previous joints. Compare the finished assembly with those made before.

REVIEW QUESTIONS

1. How does the weld made with DCRP compare with that made with HFAC?

2. How does the end of the electrode used with DCRP compare with that used with HFAC?

3. How do the observations made in this unit compare with those made in unit 5?

4. What disadvantage is there to welding light-gage aluminum without a jig?

5. What happens when the tungsten comes in contact with the work?

WELDING MAGNESIUM AND MAGNESIUM ALLOYS

The welding of magnesium and its alloys started laboratory experimentation and investigation of the use of inert-gas shielded-arc welding in the early 1940s.

CHARACTERISTICS OF MAGNESIUM

The metal is silvery white in appearance, relatively light in weight and comparatively strong. The specific gravity of magnesium is 1.74 as compared with 2.70 for aluminum. This is only 65 percent or approximately two-thirds of the weight of aluminum. Comparing this 1.74 with steel which has a specific gravity of 7.86, it is found that magnesium weighs only 22 percent or roughly one-quarter as much as steel. The melting point of magnesium is almost identical with that of pure aluminum: magnesium melts at 1204 degrees F. and aluminum, at 1218 degrees F. While magnesium itself is not too strong, its alloys possess excellent strength. The thermal conductivity of magnesium is reasonably high when compared to other metals.

WELDING MAGNESIUM

Welding magnesium is no more difficult than welding aluminum. (See chart 14-1.) The same equipment and techniques are used and the same preparation and precautions must be carried out. The rate at which magnesium expands when heated is higher than aluminum. More severe warpage results if proper precautions are not taken, especially in the lighter thicknesses.

Chart 14-1

MAGNESIUM ALLOY WELDABILITY			
Wrought Alloy	TIG Weld Rating	Cast Alloy	TIG Weld Rating
AZ10A	A	AM100A	B+
AZ31B	A	AZ63A	C
		AZ81A	B+
AZ80A	B	AZ91C	B+
HK31A	A	AZ92A	B
HM21A	A	EK41A	B
HM31A	A	EZ33A	A
LA141A	B	HK31A	B+
M1A	A	HZ32A	C
ZE10A	A	K1A	A
ZK21A	B	QE22A	B
		ZE41A	C
		ZH62A	C-
		ZK51A	D
		ZK61A	D

Welding Ratings:
A-Excellent B-Good C-Fair D-Welding not recommended

The American Welding Society

Chart 14-2

DATA FOR TIG WELDING MAGNESIUM

Material Thickness — inches	Number of Passes	Welding Current in amperes		Electrode diameter — inches			Welding Rod Diameter — inches	Gas Flow — Cu. Ft./Hr.	
		AZ31B	HK31A HM21A	HFAC	DCRP	DCSP		Argon	Helium
0.040	1	35	40	1/16	5/32	.040	3/32	12	24
0.063	1	50	55	1/16	3/16	.040	3/32	12	24
0.80	1	65	70	1/16	3/16	1/16	3/32	12	24
0.100	1	85	95	3/32	1/4	1/16	1/8	18	30
0.125	1	100	110	1/8	1/4	1/16	1/8	18	36
0.190	1	140	155	1/8	—	3/32	5/32	18	36
0.190	2	100	110	1/8	1/4	1/16	1/8	18	36
0.250	1	180	200	5/32	—	3/32	5/32	18	48
0.250	2	115	125	1/8	1/4	1/16	1/8	18	36
0.375	1	250	275	3/16	—	1/8	3/16	24	48
0.375	2	140	155	1/8	—	3/32	5/32	24	48
0.500	2	310	340	1/4	—	1/8	3/16	24	48
0.750 and over	2	350	385	1/4	—	5/32	3/16	36	48

Current values given for welding with a backing plate and for making fillet welds. Slightly lower current values used for welding without a backing plate and for making corner or edge joints.

The same types of joints and joint preparation used in welding aluminum apply here so the description is not repeated. In general, the joints are tightly butted together or, if a gap is left, it is usually less than 1/16 inch wide.

For manual TIG welding, chart 14-2 supplies the necessary data with the exception of the gas nozzles. The formula which requires that the nozzle be five to six times the diameter of the electrode applies equally in this case except for DCRP in which the nozzle should match the electrodes used for HFAC or be slightly larger. The small nozzles tend to increase the velocity of shielding gas and cause turbulence in the weld and, on occasion, draw air into the shielding zone.

Magnesium alloys are usually supplied with the surface etched to remove impurities and then oiled to preserve this surface. The oil must be removed and the surface either mechanically cleaned with abrasives or chemically cleaned prior to welding to eliminate any harmful inclusions.

SAFETY

All the standard protective equipment used in arc welding should also be used with magnesium welding. Magnesium is not particularly a fire hazard except in the form of chips, turnings, filings or other small particles. These should never be in the welding area.

One other safety precaution is necessary when welding magnesium-thorium alloys. Thorium is a radioactive element and can be toxic. A suggested limit of 0.1 mg of thorium per cubic meter of air (about 30 cubic feet of air) is a safe limit in the welder's breathing zone.

REVIEW QUESTIONS

1. What precautions are necessary when welding the magnesium-zinc alloys?

2. Why is the thermal conductivity rating of an alloy important to a welder?

3. What advantage is there in using DCRP when welding magnesium?

4. What can happen if magnesium grinding dust or particles come in contact with the welding arc?

5. On thick sections of magnesium, what might be necessary before welding?

TIG WELDING STAINLESS AND MILD STEEL

Stainless steel is one of the most widely used alloys in our modern day living. It is found in the home in tableware, kitchenware and equipment as well as in some decorative pieces. It is equally important but not as obvious in the form of heating elements in toasters, grills, electric stoves, space and water heaters and many useful applications. Stainless steel is used in hotels, restaurants, transportation, communications, paper making, and chemical and food processing industries. It is supplied in sheets, strip, plate, structural shapes, tubing, pipe, wire extrusions and in a wide variety of alloys and finishes.

In general, a welder who has mastered the TIG process in the preceding units should have no difficulty in producing welds of excellent quality and appearance in stainless and mild steel. The absence of flux and slag, the high degree of visibility in the weld area, the concentration of heat in a very narrow zone and the elimination of splatter, all contribute to produce strong, sound, smooth welds at a high rate of travel. However, the net results are equal to the degree of attention given to all the factors involved in producing TIG welds in stainless steel. The techniques of TIG welding mild steel parallel those for the welding of stainless steel very closely.

PROPERTIES AND COMPOSITION OF STAINLESS STEEL

All stainless steels derive their stainless characteristics from the chromium content. The other chief alloying element is nickel, which is added to reduce the tendency of the metal to harden as well as to give other desirable characteristics. The major producers of stainless steel can furnish details and specifications for any type or grade they produce.

WELDING STAINLESS STEEL

The actual welding of stainless steel is not difficult. However, like other metals and alloys, it has certain characteristics which must be recognized and compensated for if satisfactory joints are to be produced by any fusion process, figure 15-1.

Fig. 15-1 TIG welding stainless steel (Reprinted from Jeffus & Johnson, *Welding: Principles & Applications,* Figure 15-16)

The thermal conductivity of the alloys is about 40 percent to 50 percent less than for carbon steel. This means that the heat is retained in the weld zone much longer.

The thermal expansion is much greater than for carbon steel — about 50 percent more. This means that the shrinkage stresses are much greater and the resulting warpage becomes more of a problem if proper jigs and fixtures are not used.

In general, the metallurgical aspects of stainless steel welds are improved if the heat is carried away from the weld zone at a rapid rate. This can be accomplished by using jigs made of copper or by using copper inserts of adequate size. The electrical resistance of stainless steel varies from six times as great as carbon steel to twelve times as great, depending on the condition of the stainless steel.

JOINT PREPARATION

In the lighter gages (up to .040 inch), the best results are obtained in a butt joint by flanging both pieces. In the intermediate gages (16 through 10 gages), a square butt joint usually gives good results. In the lighter gages, a closed square butt joint is usually used. In the heavier gages, an open butt joint gives better penetration and fusion. An opening equal to one-third to one times the thickness of the material is recommended. The exact amount varies with the thickness of the material, the amount of current used, and the rate of travel. The operator, when preparing any joint for welding, must be very critical of any evidence of dirt, oil, grease, moisture or any material which might affect the finished weld.

CHOICE OF CURRENT

Generally DCSP is recommended when welding stainless steel or mild steel. In this case the cleaning action of HFAC is not necessary. It also results in a higher input of heat to the work, as discovered in the experiments in unit 5. The net result is greater penetration and higher welding speeds.

In welding the thinner sections of stainless and mild steel, the use of HFAC is recommended because of the lower heat input. The tendency to burn through or pierce the work is reduced. In the case of very light sections, the use of DCRP further reduces the tendency to burn through, but the operator must be aware of the hazards found in this type of TIG welding. Small electrodes are consumed at a rapid rate. The general rule for DCRP is that the electrode must be at least four times as great in diameter as for DCSP for a given current value. This must be rigidly followed in order to produce sound joints economically. For mild steel, use high frequency on start only.

CHOICE OF ELECTRODES, CURRENT VALUES AND SHIELDING GAS FLOW

Chart 15-1 gives the electrode size, current value and argon gas flow for a variety of sheet and plate thicknessess. For mild steel, use an argon gas flow of 20 cubic feet per hour. A 2 percent thoriated tungsten electrode is most desirable. For the smaller thicknesses of mild steel, the tungsten electrode should be sharpened to a point. The gas nozzle should be of sufficient size to insure complete shielding of the welding zone with a low velocity flow of the shielding gas.

Chart 15-1

DATA FOR MANUAL WELDING OF STAINLESS STEEL

Thickness in Inches	Type of Weld	DCSP Welding Current Flat* Amperes	Tungsten or Thoriated Electrode	Welding Speed Inches per minute	Filler Rod Diameter	Recommended Argon Flow Cu. Ft. per Hr.
1/16	Butt	80-100	1/16	12	1/16	11
	Lap	100-120	1/16	10	1/16	11
	Corner	80-100	1/16	12	1/16	11
	Fillet	90-100	1/16	10	1/16	11
3/32	Butt	100-120	1/16	12	3/32	11
	Lap	110-130	1/16	10	3/32	11
	Corner	100-120	1/16	12	3/32	11
	Fillet	110-130	1/16	10	3/32	11
1/8	Butt	120-140	1/16	12	3/32	11
	Lap	130-150	1/16	10	3/32	11
	Corner	120-140	1/16	12	3/32	11
	Fillet	130-150	1/16	10	3/32	11
3/16	Butt	200-250	3/32	10	1/8	13
	Lap	225-275	1/8	8	1/8	13
	Corner	200-250	3/32	10	1/8	13
	Fillet	225-275	1/8	8	1/8	13
1/4	Butt	275-350**	1/8	5	3/16	13
	Lap	300-375**	1/8	5	3/16	13
	Corner	275-350**	1/8	5	3/16	13
	Fillet	300-375**	1/8	5	3/16	13
1/2	Butt	350-450**	3/16	3	1/4	15
	Lap	375-475**	3/16	3	1/4	15
	Corner	375-475**	3/16	3	1/4	15

* — Current values are for flat position only. Reduce the above figures by 10% − 20% for vertical and overhead welds.

** — For current values over 250 amps., use a torch with a water-cooled nozzle.

JIGS AND FIXTURES

The use of jigs and fixtures to clamp and align the work is strongly recommended when welding stainless steel. This material has a relatively high coefficient of thermal expansion. A good jig helps to hold the work in good alignment and eliminate some of the warpage due to thermal stresses.

The thermal conductivity of stainless steel is relatively low and rapid removal of heat from the work results, in general, in a finished product with better physical and metallurgical properties. This rapid removal of heat can best be accomplished by using clamping jigs made of copper, which has a high rate of thermal conductivity, or by using jigs with copper inserts. Figures 15-2 and 15-3 show two types of steel jigs with copper inserts.

Fig. 15-2 Copper insert jig for butt welds

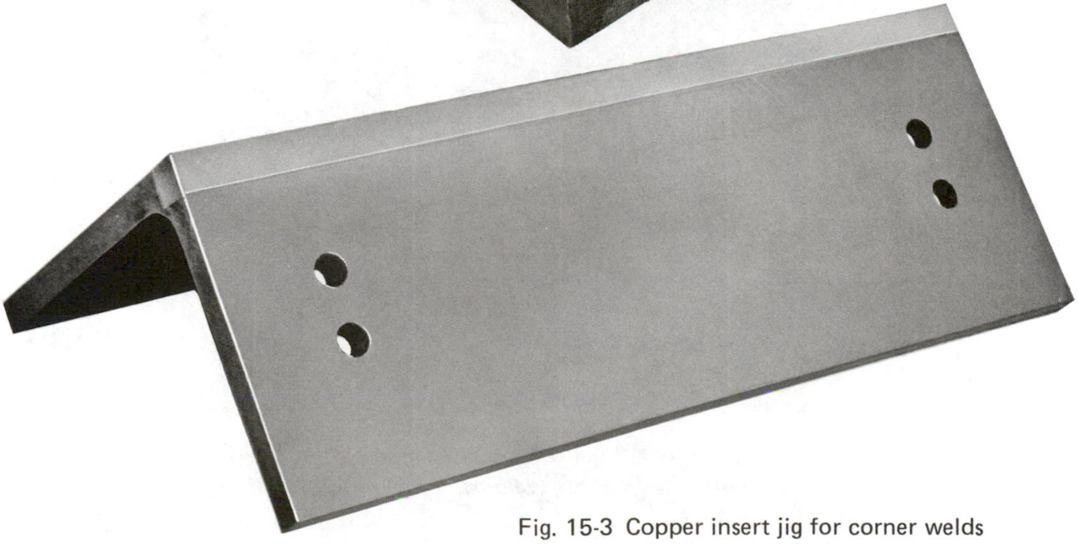

Fig. 15-3 Copper insert jig for corner welds

TECHNIQUES AND PRECAUTIONS

The gas nozzle should be as close as possible to the work consistent with good visibility. When welding with DCSP, the best results are obtained if thoriated tungsten electrodes are used. Welding in a still atmosphere free from drafts insures maximum effectiveness of the shielding gas used.

Argon gas shielding gives the smoothest arc action and a resulting smooth weld. Helium gas shielding produces a somewhat hotter arc and contributes to higher welding speeds. One of the large steel companies recommends experimenting with varying mixtures of these two gases to get the best results. In general, argon shielding gives good results when welding with all the alloys, especially in the thinner gages.

Electrode contamination must be guarded against to insure proper current flow. Accidental dipping of the electrode into the molten pool contaminates the electrode and greatly reduces the current-carrying capacity.

When filler rods are necessary, the use of a rod of the same composition as the base metal being welded gives good results, figure 15-4. The TIG process does not reduce the percentage of the alloying elements to any great extent.

PRACTICE WELDS

At this point, the student should have developed enough skill and judgment to make a series of practice welds using various thicknesses of metal to make the standard types of joints.

It is found that 10-, 12-, and 14-gage stainless steel can be welded at a slow enough rate to permit the observation of the effects of varying arc length, electrode angles and filler-rod angles. While these thicknesses are welded with little difficulty, it is advisable to use workpieces from .040 inch to 1/16 inch thick and to practice until becoming proficient in producing welds in these light gages.

Some experimenting in dipping the electrode in the molten pool and observing the arc action and the weld produced helps to point out the undesirable characteristics of electrode

Fig. 15-4 Proper filler rod results in clean welds. (Reprinted from Jeffus & Johnson, *Welding: Principles & Applications,* Figure 15-2)

contamination when welding in stainless steel. On both stainless and mild steel, the bead width should not be greater than 1/8 inch on light-gage sheet. Discoloration at each side of the weld should not extend more than twice the width of the bead, figure 15-5.

If polishing and buffing equipment is available, some of the joints made should be polished, buffed and inspected. This is done to determine if the joints are of a quality which will yield the desired finish with a minimum of finishing time.

Fig. 15-5 Properly made weld

REVIEW QUESTIONS

1. Why are jigs and fixtures more necessary when welding stainless steels?

2. How does the flow of shielding gas compare with that used for TIG welding aluminum and magnesium? Why?

3. If the finished bead is to be bright and shiny over its entire length, what procedure would be used?

4. How does the electrode extension and arc length compare with that used when welding other types of metals?

5. What is the best filler rod angle to use when TIG welding?

6. What are two reasons why DCSP is recommended for welding stainless steel?

WELDING COPPER AND COPPER-BASE ALLOYS

Copper and its alloys are among the earliest metals used by man. Its low melting point, 1981 degrees F., plus the ease with which it could be refined from its ores made it readily available. Since copper could be worked and formed easily, it was a desirable material for many articles.

The use of copper and its hundreds of alloys is so extensive today that it would be hard to visualize present-day living without these valuable materials. From the welder's viewpoint, the ability to join copper and its alloys by one or several of the welding, brazing or soldering methods is an important factor.

DESIRABLE CHARACTERISTICS OF COPPER

- Copper and its alloys are highly resistant to many forms of corrosion.

- Copper and its alloys can be fabricated and formed by all standard methods.

- Copper is one of the best conductors of electricity and of heat energy.

- Many elements can be combined with copper to form a wide variety of alloys, each with specific characteristics. For example, some of these alloys are made to be highly ductile so that they can be drawn or spun. Other alloys may be highly wear-resistant and are used for many types of bearings.

- Some of the alloys are highly resistant to fatigue and are used to make corrosion-resistant springs. When alloyed with beryllium, the resulting material can be cold worked and heat treated to produce hard tools such as hammers and cold chisels. These are useful in areas where other materials might cause dangerous sparks.

UNDESIRABLE CHARACTERISTICS OF COPPER

- Copper and many of its alloys are susceptible to a condition known as hot shortness. This means that the material becomes brittle at high temperatures. Therefore, it can present many of the same problems as aluminum does in welding.

- Copper and many of its alloys owe much of their strength to the fact that they have been cold worked. Welding operations heat the metal in the weld zone to the point where it becomes annealed and loses much of its strength.

- While pure copper possesses excellent electrical and thermal conductivity, the addition of any elements, either deliberately to form alloys, or accidentally in the form of oxides, causes a sharp decrease in both electrical and thermal conductivity. It is usually found that pure metals possess better electrical and thermal conductivity than any of their alloys.

- Repeated stresses applied to copper, such as cold-forming operations or stress reversals caused by the normal operation, can cause the metal to become increasingly hard and brittle. This can lead to rupture or a fatigue type of failure.

Chart 16-1

SUGGESTED PROCEDURES FOR INERT-GAS TUNGSTEN-ARC WELDING COPPER AND EVERDUR®						
Base Metal Thickness Inch	Weld Groove	No. of Beads	Bead No.	Filler Rod Diameter Inch	DCSP Welding Current Ranges, Amperes	
			On Backing Bar		COPPER	EVERDUR®
1/16	Square	1		3/32	150-250	80-120
3/32	Square	1		1/8	180-300	100-150
1/8	Square	1		1/8	200-350	100-200
			Without Backing			
1/8	Square	2		1/8	150-300	100-150
5/32	Square	2		1/8	150-300	100-150
3/16	Square	2		1/8	180-350	100-200
3/16	Single-vee	2	1	3/16	150-300	100-200
			Root	1/8	200-300	150-200
1/4	Single-vee	3	1	1/8	200-350	100-150
			2	3/16	200-350	150-200
			Root	1/8	200-350	150-200
3/8	Single-vee	4	1	1/8	200-350	100-200
			2	3/16	250-350	150-200
			3	1/4	350-500	150-250
			Root	1/8	250-350	150-200
1/2	Single-vee	5	1	1/8	300-450	100-200
			2	3/16	300-450	150-250
			3	1/4	350-500	150-250
			4	1/4	350-500	200-300
			Root	1/8	300-450	150-200
3/4	Double-vee	6	1, 2	1/8	300-400	100-250
			3, 4	3/16	300-450	150-300
			5, 6	1/4	350-550	200-350
1	Double-vee	8	1, 2	1/8	300-400	100-250
			3, 4	3/16	300-450	150-300
			5, 6	1/4	350-550	200-350
			7, 8	1/4	350-600	200-400

CURRENT REQUIREMENTS FOR WELDING COPPER

A study of chart 16-1 and a comparison with charts 6-2, 9-1, 10-1 and 11-1 indicate that copper requires current settings of 50 percent to 75 percent higher than equal sections of aluminum. As an example: 1/8-inch aluminum requires currents up to 160 amperes; 1/8-inch copper requires 200 amperes to 350 amperes. For 1/8-inch tungsten electrodes using straight polarity, 250 amperes is the top current limit for TIG torches with ceramic nozzles.

WELDABILITY

Pure copper presents no problems. It is originally supplied in a clean state. However, the material should still be given a thorough cleaning just before welding. The presence of oxides on the surface is indicated by the color which ranges from light green to black. No welding should ever be attempted until these oxides have been removed either chemically or mechanically.

Atmospheric conditions can affect the finished joint in copper. With some types of copper welding, a tendency to porosity in the joint increases in direct proportion to the increase in relative humidity. TIG welding offers a decided advantage by providing an ideal atmosphere for the welding process.

The copper-silicon alloys are generally readily weldable. Some of these alloys are known by their trade names such as Everdur®, Herculoy®, and Olympic®. The copper-nickel alloys, such as super-nickel and cupro-nickel, are also readily welded using the TIG process if they are carefully cleaned before welding. The copper-aluminum alloys, known as aluminum-bronze, can be welded by the TIG process if the welder uses as much care as for welding aluminum. The copper-phosphorous alloys, usually referred to as phosphor-bronze, can also be welded economically with the TIG process.

While some of its alloys present no difficulties to the welder, many of the copper alloys are not as weldable. The group of copper-zinc, copper-tin, and copper-lead alloys and combinations of the four elements are either difficult or impossible to weld by the TIG process. The difficulty comes from the tendency of the zinc, tin or lead to vaporize under the intense heat of the arc. As the percentage of any of these three elements is increased in a copper-base alloy, the probability of making an acceptable TIG weld decreases. However, the low-temperature brazing process of oxyacetylene welding may be used to join many of these otherwise unweldable alloys. The many low-temperature brazing alloys available produce joints of high strength, excellent appearance and, in many cases, good color match when it is an important factor. Chart 16-2 provides an opportunity to compare copper and some of its alloys as to strength, ductility, melting point, and electrical and thermal conductivity.

JIGS AND FIXTURES

Copper and most of its alloys present unique problems when jigs and fixtures are necessary. Steel would make an excellent material for constructing jigs, but if it becomes too hot, copper and many of its alloys tend to bond or braze to it.

Carbon or carbon inserts eliminate this tendency but the material is brittle and wears rapidly. If the arc is allowed to strike the carbon and the tungsten electrode is negative, much carbon is transferred to the tungsten (just as it was transferred from positive to negative in the experiments in unit 5). This contaminates the tungsten and greatly lowers its current-carrying characteristics.

The best material for backing bars in jigs and fixtures appears to be copper itself, if it is thick enough so that it does not melt and fuse with the material being welded. Stainless steel is also satisfactory as a backing material if it has been heated until the surface is oxidized. This oxide is very hard to remove and prevents any bonding of the copper or copper-base alloys. It has the advantage of low thermal conductivity, thus more heat can be used in

Chart 16-2

PHYSICAL CHARACTERISTICS OF COPPER AND SOME OF ITS ALLOYS

Material	Composition	Average Tensile Strength k.s.i.		% Elongation in 2 inches (annealed)	Melting Point Degrees F.	Conductivity	
		Annealed	Hard			Electrical	Thermal
Copper (Tough pitch, Electrolytic Lake)	99.9 Copper .03-.07 Oxygen	30,000 to 40,000	40,000 to 67,000	35	1981	100+	.92
Deoxidized Copper	0-10 Phos. 0.25 Si Bal. Copper	30,000 to 35,000	40,000 to 50,000	35	1981	80	.80
Common Brasses Brazing, Spring, Cartridge, Yellow Brass	21-37 Zinc 0-4 Lead 0-2 Tin Bal. Copper	30,000 to 48,000	55,000 to 100,000	45 to 15	1823 to 1634	28 to 20	.31 to .22
Naval Brass, Tobin Bronze and Muntz Metal	37-43 Zinc 0-1½ Tin 0-2 Mn 0-1½ Iron 0-13 Ni 0-2 Lead	45,000 to 60,000	50,000 to 80,000	50 to 25	1742 to 1598	26 to 6	.28 to .08
Phosphor-Bronze Bearing Bronze and Gun Metal	1-30 Tin 0-4 Zinc 0-15 Lead 0-50 P Bal. Copper	30,000 to 60,000	60,000 to 150,000	50 to 15	1967 to 1418	45 to 8	.55 to .09
Copper-Silicon, Everdur® and Herculoy®	.25-5 Sil. 0-1½ Mn 0-5 Zinc 0-2.5 Iron Bal. Copper	40,000 to 60,000	65,000 to 145,000	75 to 20	1931 to 1832	12 to 4.5	.13 to .05
Super-Nickel and Cupro-Nickel	2-30 Nickel Trace Mn Bal. Copper	35,000 to 55,000	45,000 to 90,000	30 to 50	2012 to 2237	35 to 4.5	.40 to .06
Aluminum-Bronze	1-11 Al 0-4 Iron 0-5 Nickel 0-2 Tin Bal. Copper	40,000 to 100,000	50,000 to 125,000	15 to 60	1967 to 1886	35 to 7	.37 to .10
Beryllium-Copper	1-2.5 Be 0-1 Nickel Bal. Copper	50,000 to 70,000	70,000 to 190,000	55 to 45	1889 to 1742	45 to 17	.50 to .22

welding process. Since the thermal expansion of stainless steel is very close to that of copper, the possibility of excessive stresses being set up in the weldment is eliminated.

PRACTICE WELDS

The practice welding in this unit will depend on the availability of copper and copper-base alloys. The welder should gain as much experience as possible in welding the intermediate gages (1/16 inch to 1/8 inch thick) to make butt, lap, fillet and corner welds. It would be well to experiment with welding some of the copper-base alloys which contain zinc, tin and lead in varying amounts. Examine the finished welds and determine the effects these elements have on the finished joint.

At this point the student should have enough experience to be able to test finished welds and to be able to draw intelligent conclusions.

REVIEW QUESTIONS

1. Why is DCSP recommended for copper and copper-base alloy welding?

2. In terms of safety, can welding copper be hazardous?

3. From welding experiences, what is the best type of current for welding aluminum-bronze?

4. From a study of charts 16-1 and 16-2, what can be determined about current requirements when welding copper-base alloys as compared to welding pure copper?

5. What kind of backing bars are best suited for use with copper?

Unit 17

TIG WELDING NICKEL AND NICKEL-BASE ALLOYS

Nickel and many of its alloys are widely used in the chemical industry and in food processing plants. Nickel is resistant to most of the alkalies and many acids. In chemical plants, nickel and its alloys are used to resist the highly corrosive effects of alkalies and the resultant contamination of the finished product. In food processing plants, nickel and its alloys also add to a high-purity product.

ADVANTAGES OF USING TIG WELDING FOR NICKEL

When welding nickel and its alloys, the inert-gas arc-welding process has some advantages. The flux used in electric arc welding of nickel and the flux used in oxyacetylene welding of nickel are no more of a problem than in the welding of steel. However, there is the possibility of flux entrapment in any metallic-arc welding process. TIG welding eliminates this possibility. The flux used for nickel and nickel alloys causes no difficulty at ordinary temperatures; but, if this flux is not thoroughly removed, it becomes a problem when the weldment is used at high temperatures. In this case it attacks the weld and adjacent metal and corrodes them rapidly. TIG welding avoids this difficulty. Splatter and coarse ripples, which are other defects caused by metallic arc welding, are not found when using the TIG process.

The TIG process does not have any advantage over metal arc welding in the matter of grain growth or grain structure. However, both of these processes are carried on at a much more rapid rate than with oxyacetylene welding. This more rapid rate of heating and cooling generally results in a much finer grain structure in the weld and the heat-affected zone. In general, if all the factors that go into the cost of the finished product are considered, TIG welding is no more expensive than any of the other fusion processes used to join nickel and its alloys.

SURFACE PREPARATION

While no welding should be done without proper attention to surface preparation and cleanliness, nickel and its alloys are particularly sensitive to many chemicals. A good rule to follow is — unless proven to be safe, all foreign materials must be considered harmful.

Nickel and its alloys are usually supplied in a clean condition. However, many of the fabrication processes may leave the surface in a contaminated condition. Mechanical or chemical cleaning must be done before welding to avoid porosity and cracks in the finished joint. Lead, sulfur, phosphorus and some low-melting alloys are particularly harmful, as is the residue from alkaline cleaners. All traces of these materials should be removed from both sides of the joint before welding.

If the fabricating processes have required the work to be heated, nickel oxide forms. It is dark-colored, hard and highly refractory, melting at a temperature of 3794 degrees F. It can be removed by pickling or by one of the mechanical processes such as sand blasting, light grinding or the use of abrasive cloth. The quality of the finished joint depends greatly on how well the oxides and other contamination have been removed from both sides of the joint.

The amount of joint preparation also relates to the thickness of the material being welded. In general, a U-groove is recommended for the heavier sections to keep the price of the finished joint as low as possible.

JIGS AND FIXTURES

Jig design and tacking procedures which work well on carbon steel usually give equally good results for nickel. Copper is recommended as the best material for backing bars.

CURRENT REQUIREMENTS

DCSP is recommended for welding nickel and its alloys. However, on thin sections where piercing or burn-through is a problem, HFAC gives the advantage of lower heat input to the work. Amperage requirements and electrode sizes are equal to, or very close to those used to weld equal sections of carbon steel.

SHIELDING GAS

Helium is preferred for most TIG welding operations on nickel. It results in a hotter arc and increased welding speeds. Argon gas is recommended for the lighter sections where burn-through is a problem. A gas flow from 8 to 30 cubic feet per hour is usually enough, depending on the thickness of the material. The gas nozzle should always be of sufficient size to supply the shielding gas to the weld zone at a low velocity.

TECHNIQUES AND PRECAUTIONS

The nozzle and gas cap should be checked often to be sure they are tight. A loose joint can cause the flow of shielding gas to act as a venturi and draw oxygen from the air into the shield, causing weld contamination.

The torch should be held as nearly vertical as possible for good vision. The International Nickel Company recommends that, when welding on flat surfaces, the torch never be held at an angle greater than 35 degrees from vertical. Their investigation indicates that a sharper angle may draw air into the shielded zone, causing contamination.

In general, the amount of electrode extension does not vary from the normal procedure; that is, the electrode extension is as small as possible consistent with good vision and ease of manipulation. The tungsten electrode should be ground and maintained as a pencil point. Thoriated or zirconium alloyed tungsten electrodes maintain this desired point better than pure tungsten.

A superimposed high-frequency arc also helps to maintain this shape by eliminating the need to touch start the arc. In this case, if the machine is equipped with a switch for Start Only and Continuous, the switch should be set on Start Only.

The soundness of the welds produced depends on the arc length. In general, a long arc produces porosity in nickel and nickel alloys. Whenever possible an arc length of not over

Chart 17-1

RECOMMENDED COMBINATIONS FOR TIG WELDING			
BASE METAL		**FILLER ROD**	
Nickel	200	Nickel	61
Monel	400	Monel	60
Monel K	500	Monel	64
Inconel	600	Inconel	62
Inconel	X-750	Inconel	69
Inconel	722	Inconel	69

1/16 inch is recommended. As with most other metals, the presence of hydrogen-producing moisture also results in porosity in the weld.

There seems to be a relation between the speed of welding and the porosity. When making welds in which no filler rod is used, such as outside corner welds and flanged butt welds, an increase in the rate of travel results in a more dense weld.

When filler metal is used, care should be taken to keep the hot end of the rod in the gas shield at all times to avoid oxidation and the resulting contamination. The electrode should not be permitted to come in contact with either the filler rod or the molten pool. Electrode contamination results in inferior joints in nickel just as in other metals.

The filler metals used for oxyacetylene welding of nickel and its alloys are not suitable for TIG welding. International Nickel Company supplies the filler rods shown in chart 17-1 in 36-inch lengths and diameters from 1/16 inch to 3/16 inch by thirty-seconds.

PRACTICE WELDS

The practice welding in this unit depends to some extent upon the availability of nickel and its alloys. Most plants which make these metals also make scrap material available to welding schools on a loan basis. If the material is available, thicknesses of about 1/8 inch give the best conditions for studying the action of the arc in and around the molten pool.

If possible, some experience should be gained in welding nickel and its alloys to steel. A large amount of this type of welding is used in lining steel tanks and containers with nickel. This is done to make corrosion-resistant containers that are strong and low in cost. Experience should also be gained in welding materials from .040 inch to .065 inch in thickness.

The operator should experiment by allowing the electrode to touch the molten pool and observing the arc action and bead in order to recognize the hazards in electrode contamination. Experimenting with various arc lengths and observing the finished beads gives the welder good experience in judging the source of defects caused by deviating from standard procedures. Testing of the finished joints by the methods used in previous units allows the operator to compare welds made in these materials with those made in other materials.

REVIEW QUESTIONS

1. When nickel is attached to steel as a liner material, what effect do variations in temperature have on the stresses set up between the two dissimilar metals?

2. How does the oxide of nickel compare with aluminum oxide?

3. What major point of dissimilarity is there between aluminum oxide and nickel oxide?

4. What is the major advantage of a U-groove over a V when welding thick sections?

5. How does the arc length that is used in this unit compare with that used in previous units?

MIG welding (GMAW)

In the metal inert-gas, shielded-arc process, a consumable electrode in the form of wire is fed from a spool through the torch, often referred to as a welding gun. As the wire passes through the contact tube in the gun, it picks up the welding current.

MIG welding (or GMAW — gas metal-arc welding) differs from TIG in that it is a semi-automatic process. TIG is manual.

An important factor in the MIG process is the high rate at which metal can be deposited. This high rate of metal deposit and high speed of welding, which are characteristic of MIG, result in minimum distortion and a narrow heat-affected zone.

THE METAL INERT-GAS WELDING PROCESS

From an operator's viewpoint, it is easier to gain skill in the MIG process than in the TIG process. The deposition rate is much faster with MIG than TIG although the same metals can be joined with both. The thickness of material to be joined is a factor in choosing the correct process.

MIG welding (often called metal inert-gas or GMAW — gas metal arc welding) is done by using a consumable *wire electrode* to maintain the arc and to provide filler metal. The wire electrode is fed through the torch or gun at a preset controlled speed. At the same time, an *inert gas* is fed through the gun into the weld zone to prevent contamination from the surrounding atmosphere.

ADVANTAGES OF MIG WELDING

- Arc visible to operator
- High welding speed
- No slag to remove
- Sound welds
- Weld in all positions

TYPES OF MIG WELDING

- *Spray-arc welding,* figure 18-1, is a high-current-range method which produces a rapid deposition of weld metal. It is effective in welding heavy-gage metals, producing deep weld penetration.

  At high currents, the arc stability improves and the arc becomes stiff. The transition point, when the current level causes the molten metal to spray, is governed by the wire type and size, and the type of inert gas used.

- *Short-arc welding,* figure 18-2, is a reduced-heat method with a pin arc for use on all common metals. It was developed for welding thin-gage metals to eliminate distortion, burn-through and spatter. This technique can be used in the welding of heavy thicknesses of metal.

- *MIG CO$_2$ (carbon-dioxide) welding* is a variation of the MIG process. Carbon dioxide is used as the shielding gas for

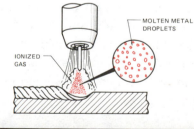

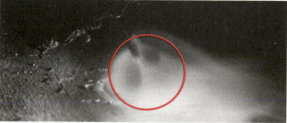

Fig. 18-1 Spray-arc welding (Reprinted from Jeffus & Johnson, *Welding: Principles & Applications,* Figure 17-9)

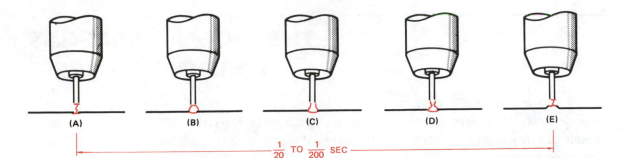

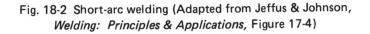

   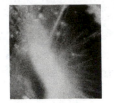 

(A) Starting with a momentary short circuit when the filler wire touches the base plate.

(B) The wire end begins to heat up.

(C) An arc is established.

(D) The wire is fed back into the molten weld pool.

(E) The arc is extinguished momentarily as the cycle starts over.

Fig. 18-2 Short-arc welding (Adapted from Jeffus & Johnson, *Welding: Principles & Applications,* Figure 17-4)

the welding of carbon and low-alloy steel from 16 gage (.059 inch) to 1/4 inch or heavier. It produces deeper penetration than argon or argon mixtures with slightly more spatter. Carbon-dioxide MIG welding costs about the same as other processes on mild steel applications.

- *Flux-cored arc welding (FCAW)* is an intense-heat, high-deposition-rate process using flux-cored wire on carbon steel. Electrically, cored-wire welding is similar to spray-arc welding. In addition to inert-gas shielding, a flux contained inside the wire forms a slag that cleans the weld and protects it from contamination. Flux-cored wires are available in diameters as small as .045 inch. The process can be used on material as thin as 1/8 inch and welded in all positions. Some flux-cored wires do not require a shielding gas.

REVIEW QUESTIONS

1. What does the term MIG welding mean?

2. What is the principle of the MIG welding process?

3. What are four types of MIG welding?

4. What polarity is used for MIG welding?

5. What are the advantages of MIG welding?

Unit 19

EQUIPMENT FOR MANUAL MIG WELDING

A specially designed welding machine is used for MIG welding. It is called a *constant-voltage (CV) type* power source. It can be a DC rectifier or a motor- or engine-driven generator or rectifier. (See figure 19-2.)

The output welding power of a CV machine has about the same voltage regardless of the welding current. The output voltage is regulated by a rheostat on the welding machine, figure 19-1. Current selection is determined by wire-feed speed. There is no current control as such.

The wire-feeding mechanism and the CV welding machine make up the heart of the MIG welding process, figure 19-3. There is a fixed relationship between the rate of electrode wire burn-off and the amount of welding current. The electrode wire-feed speed rate determines the welding current.

The gun is used to carry the electrode wire, the welding current, and the shielding gas from the wire feeder to the arc area, figure 19-4. The operator directs the arc and controls the weld with the welding gun.

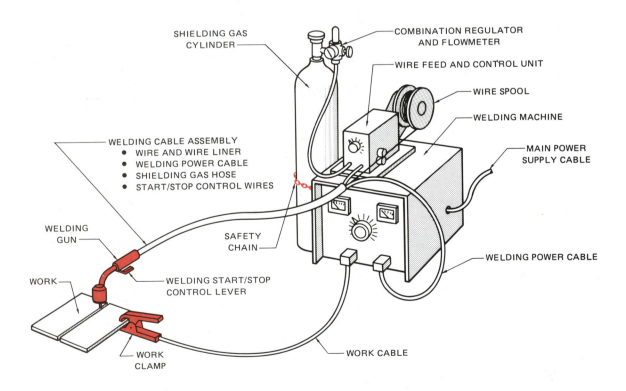

Fig. 19-1 MIG equipment (Reprinted from Jeffus & Johnson, *Welding: Principles & Applications,* Figure 1-7)

Fig. 19-2 Constant-voltage rectifier

Fig. 19-3 Wire-feed control unit

Fig. 19-4 Welding gun assembly

SHIELDING GASES

The shielding gas can have a big effect upon the properties of a weld deposit. The welding is done in a controlled atmosphere.

Pure argon, argon-helium, argon-oxygen, argon-carbon dioxide, and carbon dioxide are commonly used with the MIG process. With each kind and thickness of metal, each gas and mixture affects the smoothness of operation, weld appearance, weld quality, and welding speed in a different way.

Gas-flow rate is very important. A pressure-reducing regulator and flowmeter are required on the gas cylinder. Flow rates vary, depending on types and thicknesses of the material and the design of the joint. At times two or more gas cylinders are connected (manifolded) together to maintain higher gas flow.

FILLER WIRES

The wire electrode varies in diameter from .030 inch to 1/8 inch. The composition of the electrode wire must be matched to the base metal being welded. In the welding of carbon steel, the wire is solid and bare except for a very thin coating on the surface to prevent rusting. It must contain deoxidizers which help to clean the weld metal and to produce sound, solid welds.

REVIEW QUESTIONS

1. What are the main components of the MIG welding equipment?

2. What is considered to be the heart of the MIG welding process?

3. What does the wire-feed control determine?

4. What is the measurement of the flowmeter which registers gas flow to control the shielding atmosphere?

5. Is the voltage controlled by the wire feeder or the welding machine?

MIG WELDING VARIABLES

Most of the welding done by all processes is on carbon steel. About 90 percent of all steel is plain carbon steel. This unit describes the welding variables in short-arc welding of 24-gage to 1/4-inch mild steel sheet or plate. The type of equipment usually found in training facilities lends itself well to these applications.

The applied techniques and end results in the MIG welding process are controlled by these variables and must be understood by the student. The variables are adjustments that are to be made to the equipment and also manipulations by the operator.

These variables can be divided into three areas.

- Preselected variables
- Primary adjustable variables
- Secondary adjustable variables

PRESELECTED VARIABLES

Preselected variables depend on the type of material being welded, the thickness of the material, the welding position, the deposition rate and the mechanical properties. These variables are

- Type of electrode wire
- Size of electrode wire
- Type of inert gas
- Inert-gas flow rate

Charts 20-1, 20-2, and 20-3 are references for the new MIG welding student. Manufacturers' recommendations also serve as a guide to be followed in these areas.

PRIMARY ADJUSTABLE VARIABLES

These control the process after preselected variables have been found. They control the penetration, bead width, bead height, arc stability, deposition rate and weld soundness. They are

- Arc voltage
- Welding current
- Travel speed

SECONDARY ADJUSTABLE VARIABLES

These variables cause changes in the primary adjustable variables which in turn cause the desired change in the bead formation. They are

- Stickout
- Nozzle angle
- Wire-feed speed

Chart 20-1

COMPARISON CHART
MILD STEEL ELECTRODES FOR MIG WELDING

MANUFACTURERS	American Welding Society Classification						A5-18-6		
	E 70S-1	E 70S-2	E 70S-3	E 70S-4	E 70S-5	E 70S-6	E 70S-G	E 70S-1B	E 70S GB
Airco Welding Products Div. Air Reduction Co. Inc.	S-20		A 675		A 666	A 681	A 608	A 608	A 608
Alloy Rods Company Div. Chemetron Corporation			MINIARC 70					MINIARC 100	
Hobart Brothers Company	TYPE 20		TYPE 25		TYPE 30	TYPE 28		TYPE 18	
Linde Div. Union Carbide Corporation	LINDE 29S	LINDE 65	LINDE 82, 66	LINDE 85		LINDE 86	LINDE 83	LINDE 83	
Midstates Steel & Wire Co.			IMPERIAL 75			IMPERIAL 88	IMPERIAL 95		
Modern Engineering Co. Inc.			MECO 60S-3		MECO 70S-5		MECO 70S-G		
Murex Welding Products			MUREX 1316		MUREX 1315		MUREX 1313 MO	MUREX 1313 MO	
National Cylinder Gas Div. Chemetron Corporation			MINIARC 70				MINIARC 100		
National Standard Company	NS-106	NS-103	NS-101			NS-115	NS-116	NS-102	
P & H Welding Products Unit of Chemetron Corporation			P & H CO-85		P & H CO-86		P & H CO-87		
Page, Division of ACCO	PAGE AS-20		PAGE AS-25		PAGE AS-30	PAGE AS-28		PAGE AS-18	

ALL JOINTS ALL POSITIONS MILD STEEL							
MATERIAL THICKNESS		NUMBER OF PASSES	WIRE DIAMETER	WELDING CONDITIONS DCRP		GAS FLOW CFH	TRAVEL SPEED IPM
GAGE	INCH			ARC VOLTS	AMPERES		
24	.023	1	.030	15-17	30-50	15-20	15-20
22	.029	1	.030	15-17	40-60	15-20	18-22
20	.035	1	.035	15-17	65-85	15-20	35-40
18	.047	1	.035	17-19	80-100	15-20	35-40
16	.059	1	.035	17-19	90-110	20-25	30-35
14	.074	1	.035	18-20	110-130	20-25	25-30
12	.104	1	.035	19-21	115-135	20-25	20-25
11	.119	1	.035	19-22	120-140	20-25	20-25
10	.134	1	.045	19-23	140-180	20-25	27-32
	3/16 in.	1	.045	19-23	180-200	20-25	18-22
	1/4 in.	1	.045	20-23	180-200	20-25	12-18

Chart 20-3

SHIELDING GASES FOR MIG		
METAL	SHIELDING GAS	APPLICATION
CARBON STEEL	75% ARGON 25% CO_2	1/8 inch or less thickness: High welding speeds without burn-through; minimum distortion and spatter
	75% ARGON 25% CO_2	1/8 inch or more thickness: Minimum spatter, good control in vertical and overhead position
	CO_2	Deeper penetration, faster welding speeds
STAINLESS STEEL	90% HELIUM 7.5% ARGON 2.5% CO_2	No effect on corrosion resistance, small heat-affected zone, no undercutting, minimum distortion
LOW ALLOY STEEL	60-70% HELIUM 25-35% ARGON 4-5% CO_2	Minimum reactivity, excellent toughness, excellent arc stability and bead contour, little spatter
	75% ARGON 25% CO_2	Fair toughness, excellent arc stability, and bead contour, little spatter
ALUMINUM, COPPER, MAGNESIUM, NICKEL AND THEIR ALLOYS	ARGON AND ARGON-HELIUM	Argon satisfactory on lighter material, Argon-helium preferred on thicker material

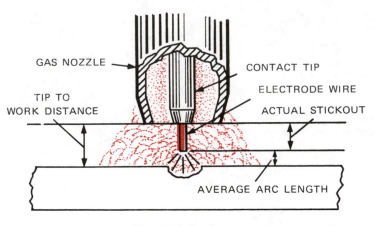

Fig. 20-1 Stickout

Stickout as shown in figure 20-1 is the distance between the end of the contact tip and the end of the electrode wire. From the operator's viewpoint, however, stickout is the distance between the end of the nozzle and the surface of the work.

Nozzle angle refers to the position of the welding gun in relation to the joint as shown in figure 20-2. The *transverse angle* is usually one-half of the included angle between plates forming the joints. The *longitudinal angle* is the angle between the centerline of the welding gun and a line perpendicular to the axis of the weld.

The longitudinal angle is generally called the nozzle angle and is shown in figure 20-3 as either trailing (pulling) or leading (pushing). Whether the operator is left-handed or right-handed has to be considered to realize the effects of each angle in relation to the direction of travel.

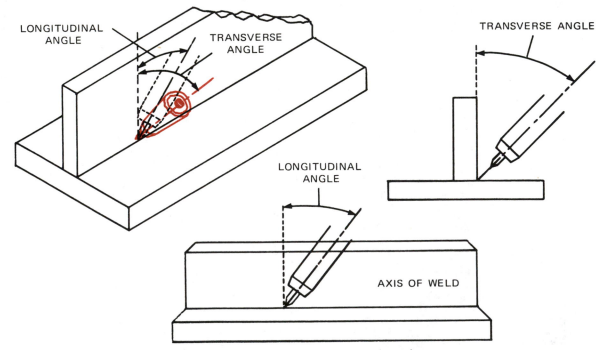

Fig. 20-2 Transverse and longitudinal nozzle angles

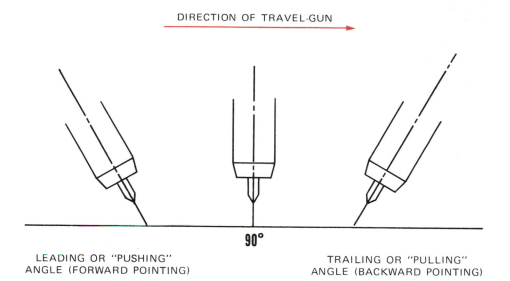

DIRECTION OF TRAVEL-GUN

90°

LEADING OR "PUSHING"
ANGLE (FORWARD POINTING)

TRAILING OR "PULLING"
ANGLE (BACKWARD POINTING)

Fig. 20-3 Nozzle angle, right-handed operator

REVIEW QUESTIONS

1. What must be considered before selecting the type and size of electrode wire and type of inert gas?

2. What controls the penetration and the bead width and height?

3. What is the distance between the end of the nozzle and the work called?

4. What are the two nozzle angles called?

5. Where does the electrode wire pick up its electrical current?

Unit 21

ESTABLISHING THE ARC AND MAKING WELD BEADS

It is assumed that the welding equipment has been set up according to procedures outlined in the appropriate manufacturers' instruction manuals. Students should know how to perform adjustments and maintenance on MIG welding equipment. As in TIG welding, the equipment is expensive, and the student must realize that the equipment can be destroyed if instructions are not followed.

Materials

> 10-, 11-, or 12-gage mild steel plate 6 in. x 6 in.
> .035-inch E 70S-3 electrode wire
> CO_2 shielding gas

Preweld Procedure

1. Check the operation manuals for manufacturer's recommendations.

2. Set the voltage at about 19 volts.

3. Set the wire-feed speed control to produce a welding current of 110 to 135 amperes.

4. Adjust the gas-flow rate to 20 cubic feet per hour.

5. Recess the contact tip from the front edge of the nozzle 0 to 1/8 inch.

6. Review standard safe practice procedures in ventilation, eye and face protection, fire, compressed gas and preventive maintenance. Safety precautions should always be part of the preweld procedure.

Welding Procedure

1. Maintain the tip-to-work distance of 3/8 inch (stickout) at all times. See figure 20-1.

2. Maintain the trailing gun transverse angle at 90 degrees and the longitudinal angle at 30 degrees from perpendicular. See figures 20-2 and 20-3.

3. Hold the gun 3/8 inch from the work, lower the helmet by shaking the head, and squeeze the trigger to start the controls and establish the arc.

    Note: Operators should not form the habit of lowering the helmet by hand since one hand must hold the gun and the other may be needed to hold pieces to be tacked or positioned.

4. Make a single downhand stringer weld.

    Note: It is permissible to travel with a steady, fixed position or a slight oscillating motion, figures 21-1 and 21-2. Figure 21-2 shows a partially completed padded plate using this motion.

5. Practice welding beads. Start at one edge and weld across the plate to the opposite edge.

Fig. 21-1 Weld bead made with axial spray metal transfer (Reprinted from Jeffus & Johnson, *Welding: Principles & Applications,* Figure 18-45)

Fig. 21-2 Partially completed pad using oscillating motion

Note: When the equipment is properly adjusted, a rapidly crackling or hissing sound of the arc is a good indicator of correct arc length.

6. Practice stopping in the middle of the plate, restarting into the existing crater and continuing the weld bead across the plate.

Note: When the gun trigger is released after welding, the electrode forms a ball on the end. To the new operator, this may present a problem in obtaining the penetration needed at the start. This can be corrected by cutting the ball off with wire cutting pliers. The ball can cause whiskers to be deposited at the start. *Whiskers* are short lengths of electrode which have not been consumed into the weld bead.

This procedure should be practiced often by the new operator. A satisfactory performance in welding joints depends on the ability to do this basic manipulation.

Checking Application

1. Examine the base metal to be sure it is free from oil, scale, and rust.

2. Recheck the equipment settings according to the operation manual. This includes gas flow, stickout, gun angle, arc voltage, and amperage from wire-feed speed.

3. Keep equipment clean, specifically the gun nozzle, feeder rolls, wire guides, and liners. An anti-spatter spray should be used on the nozzle to help keep it clean.

REVIEW QUESTIONS

1. What is a good indication of correct arc length when the equipment is adjusted properly?

2. Why is it necessary to have the stickout distance correct, especially at the start and end of the weld?

3. Why is spatter buildup on the inside of the nozzle harmful to a good weld?

4. What is a whisker in MIG welding?

5. Why is the ball end cut off the wire?

MIG WELDING THE BASIC JOINTS

The ability to manipulate the equipment and apply the single bead across a piece of sheet or plate is the basis for the welding of various joints. Being able to see and follow a joint helps to insure equal fusion on both pieces.

Note: Depending on the size of the gun, visibility may be a problem. The operator should be in the most comfortable position to see where the deposit is being made. Visibility is usually better using a leading gun angle, but penetration is greater. This could present a problem on sheet metal.

The welds that make up these basic joints can be applied in any position and can be single or multiple pass, depending on the thickness of the material being joined.

This unit describes the basics involved in welding the butt joint, lap joint, T joint, and corner joint in the flat position using 11-gage mild steel sheet. It is assumed that the operator can position and tack two pieces of this material to form these joints.

Materials

10-, 11-, or 12-gage mild steel sheet 1 1/2 in. x 6 in.
.035-inch E 70S-3 electrode wire
CO_2 shielding gas

Preweld Procedure

1. Check the operation manual for the manufacturers' recommendations.

2. Set the voltage at about 21 volts.

3. Set the wire-feed speed control to produce a welding current of 110 to 135 amperes.

4. Adjust the gas flow to 20 cubic feet per hour.

5. Adjust the voltage to get a smooth arc.

    Note: Before attempting to weld a joint, always adjust the machine, using a piece of scrap material.

6. Recess the contact tip from the edge of the nozzle 0 to 1/8 inch.

7. Make sure that the equipment is clean; specifically the gun nozzle and liner, feeder rolls and wire guides.

8. Remember that safety should always be part of the preweld procedure.

Welding Procedure (Butt Weld)

1. Place two pieces on the worktable in good alignment and with two 6-inch edges spaced 1/16 inch apart (root opening), as in figure 22-1.

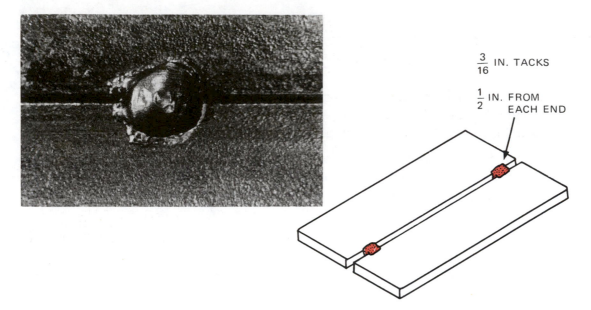

$\frac{3}{16}$ IN. TACKS

$\frac{1}{2}$ IN. FROM EACH END

Fig. 22-1 Butt joint (Photo adapted from Jeffus & Johnson, *Welding: Principles & Applications,* Figure 18-24)

2. Maintain a stickout of 3/8 inch. See figure 20-1.

3. Tack weld the two pieces together as shown in figure 22-1. The tacks should be placed about 1/2 inch from each end to avoid having too much metal and poor penetration at the start of the weld.

   Note: This method of tacking is used in all of the joints.

4. Use a transverse angle of 90 degrees or directly over and centered on the joint. Find the longitudinal angle by experimentation. The trailing gun angle is used first at 10 degrees perpendicular. The leading gun angle is used next at the same inclination. See figures 20-2 and 20-3.

   Note: Differences should be found and allowed for. The operator will prefer either the trailing gun angle or the leading gun angle, but the deciding factors are the degree of penetration, deposition rate, gas coverage of weld zone, and overall appearance.

5. Weld the joint on the tack side.

6. Cool the work and examine for uniformity.

7. Check the depth of penetration of the weld by first placing the assembly in a vise with the center of the weld slightly above and parallel to the jaws. Then bend the outstanding sheet toward the face of the weld. Penetration should be 100 percent with no faults.

8. Make more joints of this type and change the setting of the equipment slightly in different directions. Note what happens and what has to be done to compensate for it.

   Note: Thickness of material controls the root opening and whether or not the butt joint will need a bead on both sides.

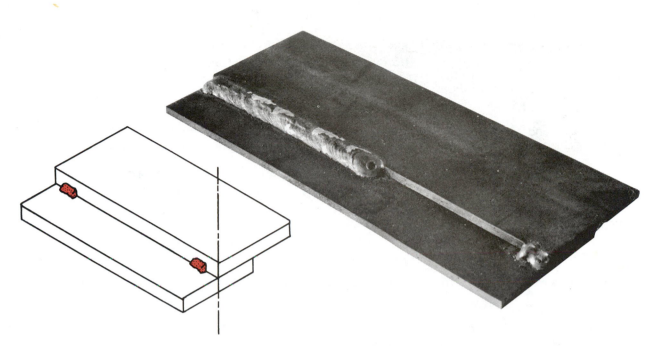

Fig. 22-2 Lap joint

Welding Procedure (Lap Joint)

1. Place two pieces on the worktable and tack as shown in figure 22-2.

2. Maintain a stickout of 3/8 inch. See figure 20-1.

3. The longitudinal angle is 10 degrees from perpendicular using a trailing gun angle. The transverse gun angle should be about 60 degrees from the lower sheet. The location of the nozzle in relation to the joint should be as shown in figure 22-3.

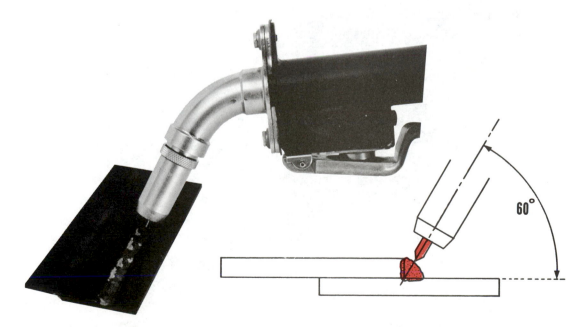

Fig. 22-3 Location of nozzle

Note: The operator's ability to compensate for the location of the gun in relation to the joint controls the uniformity of the bead and the desired amount of penetration.

4. Weld the joint. Allow for distortion by running the first bead on the opposite side from the tacks.

5. Cool and examine the bead for uniformity. Examine the line of fusion with the top and bottom sheets. This should be a straight line with no undercut.

6. Weld another lap joint on only one side (tack side). Place this piece in a vise in such a way that the top sheet can be bent from the bottom sheet 180 degrees, if possible, to check penetration and strength.

7. Make another test by sawing a lap-welded specimen in two and examining the cross section for penetration.

8. Make more joints of this type until they are uniform and consistent.

Welding Procedure (T Joint)

1. Place two pieces on the worktable and tack weld as shown in figure 22-4.

   Note: Hold one piece while tacking. This is good experience, as it is necessary to handle and manipulate the gun with one hand.

2. Maintain a stickout of 3/8 inch. See figure 20-1.

Fig. 22-4 T joint

3. The longitudinal angle is 10 degrees from perpendicular using a trailing gun angle. The transverse gun angle should be about 45 degrees from the lower piece. The location of the nozzle in relation to the joint should be as shown in figure 22-5.

4. Weld the joint. Allow for distortion by running the first bead on the opposite side from the tacks.

5. Cool and examine the bead for uniformity. The weld metal should be equally distributed between both pieces and show no signs of undercut.

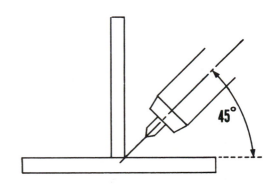

Fig. 22-5 Transverse angle

Note: A tack weld should be strong enough to resist cracking during the welding process but not large enough to affect the appearance of the finished weld. This is done in MIG welding by using slightly higher wire-feed speed.

6. Make another T joint welding only the tack side. Test this weld by bending the top piece against the joint a full 90 degrees. Examine the joint for root penetration and uniform fusion.

7. Continue to make the fillet weld until acceptable welds can be made each time. The fillet-type weld used on the lap and T joint is the most common weld.

Welding Procedure (Corner Joint)

Note: The technique used to set up and align the pieces to be joined for the down-hand corner joint is more difficult. Figure 22-6 shows how this is done without using a fixture.

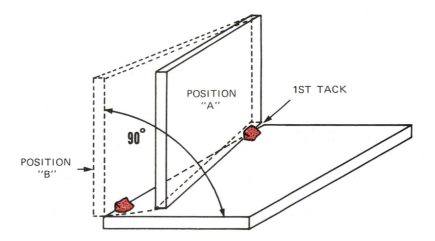

Fig. 22-6 Setting up for corner joint

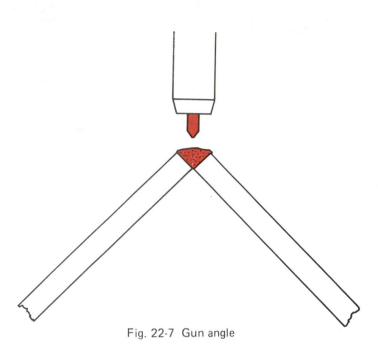

Fig. 22-7 Gun angle

1. Make the first tack while holding the piece as shown by position A, figure 22-6. Then lift the hood and align the top piece to position B. Make a perfect open corner joint before placement of the second tack.

2. Maintain a stickout of 1/4 inch to 3/8 inch.

3. The longitudinal angle is 10 degrees from perpendicular using a trailing gun angle. The transverse gun angle should be perpendicular or bisect the included angle. See figure 22-7.

4. Weld the joint. Pay close attention to the start and end of the joint to avoid buildup or washout.

5. Cool and examine the bead for uniformity and penetration. The weld metal should be equally distributed between both pieces and show no signs of undercut or overlap. See figure 22-8.

Fig. 22-8 Examples of uniform welds

6. To test the corner joint, place the welded unit on an anvil and hammer it flat in order to examine root fusion and penetration.

7. Make more corner joints until they have uniform appearance and a good finish contour. The opposite side of this joint provides for good fillet weld practice.

Checking Application

1. Recheck the equipment settings according to the operation manual.

2. Keep the equipment clean.

3. Practice these four basic joints in the flat position using thicknesses of material up through 3/16 inch.

REVIEW QUESTIONS

1. When a tack is placed on two pieces of material being joined, what is the function of the tack?

2. What are the two types of nozzle angles as related to the longitudinal angle used in the operation of the gun?

3. What causes undercut and why is it harmful to the strength of the weld?

4. When welding any joint, what is important concerning bead location?

5. Of the four joints, butt, lap, corner and T, which one might require more inert gas? Why?

Unit 23

PROCEDURE VARIABLES

OUT-OF-POSITION WELDING

Upon satisfactory completion of the welds in the flat position, the student will be able to use the acquired skill and knowledge to weld out of position. This includes horizontal, vertical-up, vertical-down, and overhead welds. The basic procedures for each individual joint are no different out of position than in the flat position except a reduction in amperage of 10 percent is usually recommended. See chart 20-2.

MISALIGNED MATERIALS

The operator may, at times, have to weld pieces of material that are not in plane or aligned properly. There may be gaps or voids of various sizes which need a variation of stickout and/or wire-feed speed and voltage. The student should practice on the joints that require a deviation from standard procedures.

WELDING HEAVIER THICKNESSES

Heavier thicknesses of material can be welded with the MIG process using the multipass technique. This is done by overlapping single small beads or progressively making larger beads, using the weave technique, as in figure 23-1. The numbers refer to the order in which the passes are made. Individual job requirements govern the end result.

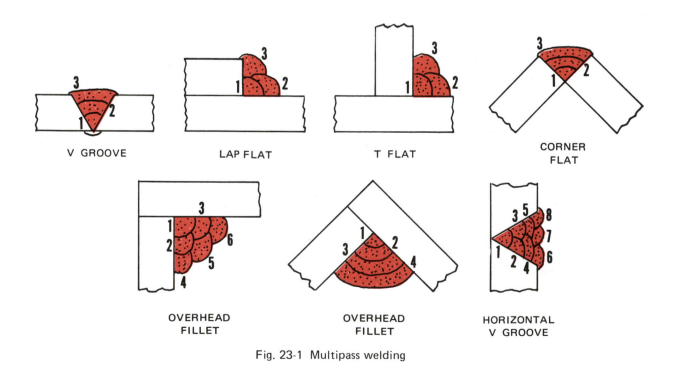

Fig. 23-1 Multipass welding

| **DEFECTS** | **PROBABLE CAUSE** | **CORRECTIVE ACTION** |

Fig. 23-2 Porosity

Gas flow does not displace air, clogged or defective system, frozen regulator — Set gas flow between 15 and 23 CFH. Clean spatter from nozzle often. Use a regulator heater when drawing over 25 CFH of CO_2

Fig. 23-3 Porosity in crater at end of weld

Pulling gun and gas shield away before crater has solidified — Reduce travel speed at end of joint

Fig. 23-4 Cold lap lack of fusion

Improper technique preventing arc from melting base metal — Direct the welding arc so that it covers all areas of the joint. Do not allow the puddle to do the fusing. Use a slight whip motion.

Fig. 23-5 Burn-through and too much penetration

Heat input too high in the weld area — Reduce wire-feed speed to obtain lower amperage. Increase travel speed. Oscillate gun slightly. Increase stickout to 1/2 inch maximum

Fig. 23-6 Lack of penetration

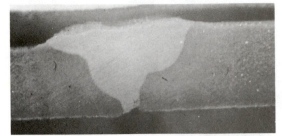

Heat input too low in the weld area

Increase wire-feed speed to obtain higher amperage. Reduce stickout to 1/4 inch.

Fig. 23-7 Whiskers

Electrode wire pushed past the front of the weld puddle leaving unmelted wire on the root side of the joint

Cut off ball on end of wire with pliers before pulling trigger. Reduce travel speed and, if necessary, use a whipping motion.

Fig. 23-8 Wagon tracks

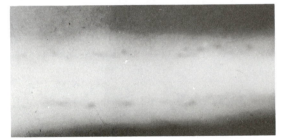

Too high bead contour or too high crown. Area where bead fuses to side of joint is depressed and next bead may not completely fill depressed area or void.

Arc voltage and travel speed should be high enough to prevent crown. When welding over these areas, be sure that the welding arc melts the underlying weld and base metal.

CAUSE AND CORRECTION OF DEFECTS

The operator has to be able to recognize and correct possible welding defects. MIG welding, like the other processes, must be properly applied and controlled to consistently give high-quality welds. The defects are shown in figures 23-2 through 23-8, accompanied by the causes and corrective actions to be taken.

REVIEW QUESTIONS

1. What positions does out-of-position welding refer to?

2. What is porosity in a MIG weld?

3. What quality will the weld probably lack if the current input is too low at the arc?

4. In MIG welding heavy plate, is more amperage and voltage required than welding light plate? Why?

5. Can the MIG gun be oscillated to improve bead conformity?

Unit 24

MIG WELDING ALUMINUM

The welding of aluminum using the MIG process is advantageous over the TIG process because heavier sections of metal can be welded much faster. Aluminum and its alloys differ from mild steel in that there is no color change as the temperature from welding increases. Aluminum also develops a refractory oxide when exposed to air. Although material of .040 inch can be MIG welded, 3/16 inch is about the minimum for spray-arc welding.

Equipment

The constant-voltage machine should have a potential output of 500 amperes. Due to this increased output the standard gun is usually water cooled. There are guns which hold one pound of smaller diameter aluminum wire which do not require water for cooling.

Materials

1/4-inch aluminum plate
1/16-inch E4043 electrode wire
One cylinder of argon gas

Preweld Procedure

1. Set the voltage at 23 to 27 volts.

2. The wire-feed speed should produce a welding current of 225 to 300 amperes.

   Note: From these basic settings, a smooth transfer of metal across the arc can be obtained with slight adjustments. Voltage, amperage, and welding techniques vary to suit the joint and position conditions.

3. Adjust the gas flow to about 35 CFH.

4. Always clean the aluminum before welding. This can be done with suitable commercial solvents or by mechanically filing, scraping or brushing with a stainless-steel wire brush.

5. Always check the equipment manufacturers' recommendations before welding.

Welding Procedure

1. Maintain the stickout from 1/2 inch to 3/4 inch on all joints.

2. Use the same transverse angle for all joints as for mild steel.

3. Use the same longitudinal angle for all joints as for mild steel, except a leading gun angle should be used. See figure 20-3.

4. Hold the electrode wire toward the leading edge of the puddle. The forward motion can be steady or oscillating, depending on the application.

5. Check the joint preparation of aluminum since it is critical and the operator must be sure of good fitup. The butt joint, lap joint, and T joint are the best joints for fabrication.

Fig. 24-1 A lap weld made by spray arc. Note the oscillation ripples in the weld.

6. Do the setup, tacking, welding and testing of the aluminum joints as for mild steel.

Note: The spray-arc type of metal transfer can be used for out-of-position welding. Because of its fluidity, however, it is more difficult than short-arc welding, figure 24-1.

Checking Application

1. The surfaces of the plate must be thoroughly clean for the best results.

2. Good joint fitup is necessary to provide weld puddle control and to prevent unnecessary distortion.

REVIEW QUESTIONS

1. What are two characteristics in welding aluminum that are different from those of mild steel?

2. What does the term oscillate mean in reference to the manipulation of the MIG gun and what is its function?

3. What type of gun angle is used in aluminum MIG welding?

4. What type of MIG welding is used on aluminum?

5. What is the minimum thickness of aluminum that can be MIG welded economically?

Unit 25

MIG WELDING STAINLESS STEEL

The techniques involved in MIG welding stainless steel are similar to those used for mild steel. Spray-arc or short-arc welding can be used in the welding of stainless steel, depending on the thickness of the material being welded and the amount of current being produced. Refer to charts 25-1 and 25-2 for the welding conditions of spray-arc and short-arc welding, respectively.

Shielding gas should be argon with 1 percent oxygen, or argon with 2 percent oxygen depending on the thickness of the material being welded. In abbreviated form this is written O_2-1 and O_2-2.

Chart 25-1

GENERAL WELDING CONDITIONS, SPRAY ARC							
Plate Thickness (In.)	Joint & Edge Preparation	Wire Dia.	Gas Flow	Current (DCRP Amps)	Wire Feed (ipm)	Welding Speed	Passes
.125	Square Butt with Backing	1/16	35	200–250	110–150	20	1
.250	Single Vee Butt 60° Inc. Angle No Nose	1/16	35	250–300	150–200	15	2
.375	Single Vee Butt 60° Inc. Angle 1/16-in. nose	1/16	35 (O_2-1)	275–325	225–250	20	2
.500	Single Vee Butt 60° Inc. Angle 1/16-in. nose	3/32	35 (O_2-1)	300–350	75–85	5	3–4
.750	Single Vee Butt 90° Inc. Angle 1/16-in. nose	3/32	35 (O_2-1)	350–375	85–95	4	5–6
1.000	Single Vee Butt 90° Welded Angle 1/16-in. nose	3/32	35 (O_2-1)	350–375	85–95	2	7–8

Chart 25-2

Plate Thickness (In.)	Joint and Edge Preparation	Wire Dia. (In.)	Gas Flow (CFH)	Current DCRP (amps)	Voltage	Wire-Feed Speed (ipm)	Welding Speed (ipm)	Passes
	GENERAL WELDING CONDITIONS, SHORT ARC							
.063	Nonpositioned fillet or lap	.035	15-20	85	15	184	18	1
.063	Butt (square edge)	.035	15-20	85	15	184	20	1
.078	Nonpositioned fillet or lap	.035	15-20	90	15	192	14	1
.078	Butt (square edge)	.035	15-20	90	15	192	12	1
.093	Nonpositioned fillet or lap	.035	15-20	105	17	232	15	1
.125	Nonpositioned fillet or lap	.035	15-20	125	17	280	16	1

A leading gun angle is used to give more visibility. An oscillating motion back and forth in the direction of the joint is desirable for fusion and uniformity.

16-gage stainless steel
.035 inch, E308 electrode wire
75% argon and 25% CO_2 gas mixture

Preweld Procedure

1. Check the operation manual for the manufacturers' recommendations.

2. Copper backup bars are required for welding stainless steel, especially when welding only one side.

Welding Procedure

1. Arrange pieces of stainless steel to form joints as detailed in unit 22.

2. Stickout should be 1/4 inch to 3/8 inch.

3. Use either a leading or a trailing gun angle. The transverse gun angle always bisects the joint.

Checking Application

1. The workpiece must be thoroughly cleaned.

2. The wire-feed speed and the voltage settings are critical. A slight variation from the correct settings could produce unsatisfactory welds.

REVIEW QUESTIONS

1. What shielding inert gas is used in the MIG welding of stainless steel?

2. What kind of material is used for backup bars on stainless steel?

3. What does the word bisect mean?

4. What two types of MIG welding are used on stainless steel?

5. How is the thickness of stainless steel sheet specified?

MIG WELDING COPPER

No other process is as good or as fast as MIG for welding copper or its alloys such as manganese-bronze, aluminum-bronze, silicon-bronze, phosphor-bronze, cupro-nickel, and some of the tin bronzes. To obtain high-quality welds in copper it is necessary to use deoxidized, non-oxygen-bearing forms of copper, copper-base material and filler material. Chart 26-1 gives the conditions for welding copper.

Argon is the preferred inert gas for welding 1/4 inch and thinner material. A flow of 50 cubic feet per hour gives a good shielding atmosphere. For greater thicknesses, a mixture of 65 percent helium and 35 percent argon offsets the high-heat conductivity.

Preweld Procedure

1. Preheat thicknesses of over 1/4 inch to 400 degrees F.

2. Use steel for backup when required on light thicknesses.

Welding Procedure

1. Arrange pieces of copper having thicknesses of at least 1/4 inch to form groove joints and T joints.

2. Use a stickout of 1/2 inch to 3/4 inch and an oscillating motion to weld the joints.

3. Use a leading gun angle for better visibility.

   Note: Good ventilation is important when welding copper and its alloys. The fumes that are produced are highly toxic and must be carried away from the operator.

Checking Application

1. Check the operation manual for manufacturers' recommendations.

2. The welding operator must be properly protected from radiation because of its intensity.

Chart 26-1

COPPER WELDING CONDITIONS						
Thickness (In.)	Amps DCRP	Volts	Travel (ipm)	Wire Dia. (In.)	Wire-Feed Speed (ipm)	Joint Design
1/8	310	27	30	1/16	200	Square butt, steel backup strip required
1/4 (1) 1/4 (2)	460 500	26	20	3/32	135 150	Square butt
3/8 (1) 3/8 (2)	500 550	27	14	3/32	150 170	Double bevel, 90° included angle, 3/16-in. nose
1/2 (1) 1/2 (2)	540 600	27	12 10	3/32	165 180	Double bevel, 90° included angle, 1/4-in. nose

REVIEW QUESTIONS

1. In terms of safety, what has to be foremost in the mind of the welder who is welding copper?

2. How many cubic feet per hour of inert gas is preferred for welding copper?

3. What kind of a backup is required on copper?

4. How much preheat is required on copper plate prior to welding?

5. What does heat conductivity mean?

FLUX-CORED CO₂ SHIELDED MIG WELDING

The cored-wire welding process is a gas-shielded metal-arc welding process which uses the intense heat of an electric arc between a cored, consumable, continuously-fed electrode wire and the work.

SHIELDING GAS

The shielding gas (CO_2) (carbon dioxide, welding grade) displaces the air surrounding the arc and the weld puddle thus preventing contamination of the weld metal by atmospheric oxygen and nitrogen. This process requires a gas-flow rate of 35 to 60 cubic feet per hour. It may be necessary at times to have two or more cylinders manifolded together.

ELECTRODE WIRES

The flux-cored electrode wires, chart 27-1, recommended for use with this process, range in size from .045 inch to 1/8 inch diameter. The fluxes present within the core of these electrode wires add deoxidizers and strengthening elements to the deposited metal. The core also contains slag-forming and arc-modifying substances which protect the weld metal from atmospheric contamination, figure 27-1.

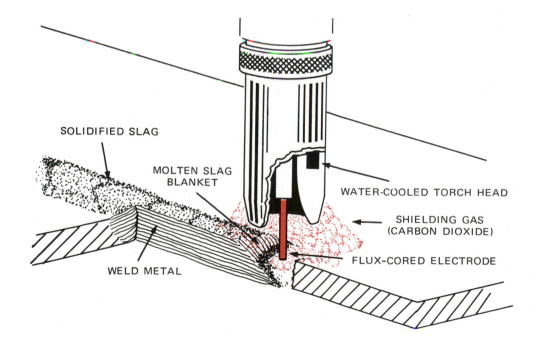

Fig. 27-1 Flux-cored gas-shielded welding

Chart 27-1

CARBON STEEL ELECTRODES — FLUX-CORED ARC WELDING

MANUFACTURER	American Welding Society Classification AWS A5-20-79								
	E 6IT-8	E 70T-1	E 70T-2	E 70T-3	E 70T-4	E 70T-5	E 70T-6	E 70T-7	E 70T-G
Air Products & Chemicals Inc.		AP 701 GP							
Airco Welding Products		Tuf-Cor Super-Cor Flux-Cor 6 Metal-Cor 6	Flux-Cor 1			Flux-Cor 5			Tensil-Cor
Alloy Rods Div.		Dual Shield R-70, 111A-C, T-62, 78, 111A, 111-HD	Dual Shield SP, T-63, 110		Coreshield 40	Dual Shield T-75			
Canadian Liquid Air Ltd.		Dual Shield 111AC LA T-9	Dual Shield SP		LA Unishield				
Canadian Rockweld Ltd.		Rockweld 72	Rockweld 70			Rockweld 80B			
Hobart Brothers Co.		FABCO 81, 82 82HD, RXR	FABCO 87	FABSHIELD 55	FABSHIELD 4	FABCO 801		FABSHIELD 31	FABCO 85
Lincoln Electric	NR-203 NICKEL C			NR-1 NR-5	NR-301 NS-3M		NR-302	NR-311	
Murex Welding Products		Mure-Cor 3, 2A, 6 MUREMET 6				Mure-Cor 5			Mure-Cor MNM
Page Welding Div. Acco Industries, Inc.		PFC-721	PFC-792						
Tri-Mark Inc.		TM-11 TM-1HE	TM-22		TM-44	TM-55			
Unicore Inc.		Uniflux 70, F77				Uniflux 75			Uniflux 100
Linde Div. Union Carbide Corp.		Linde FC-72							

APPLICATION

- Due to the very fluid weld puddle, satisfactory welds are best obtained by welding in the flat and horizontal positions.

- This process is used mainly for welding low- and medium-carbon steels and low-alloy, high-strength steels in thicknesses above 3/16 inch.

- Thicknesses up to about 1/2 inch are weldable with no edge preparation, and above 1/2 inch with edge preparation. Maximum thickness is unlimited.

- The electrode wires are designed for either single-pass or multipass procedure.

Welding Procedure

1. Set the stickout distance at 1 inch. It can be varied somewhat to control the weld.

2. Use a slightly leading gun angle. See figure 20-3.

3. Place the voltage setting between 25 and 36 volts. The wire-feed speed should produce from 250 to 700 amperes, depending on the application.

4. Using 1/2-inch plate, set up the lap joint and the T joint and weld according to preceding information.

Checking Application

1. Weld bead should have good conformity and penetration.

   Note: Penetration can be checked by cutting a specimen with a cutting torch through the weld and then grinding and filing the surface.

2. There should be no evidence of undercut.

REVIEW QUESTIONS

1. In the cored-wire welding process, how is the electrode wire different from standard MIG welding?

2. How does the deposit from cored wire compare to that of solid wire?

3. What kind of gun angle is used in cored welding?

4. What does the symbol CO_2 stand for?

5. How is the weld protected from atmospheric contamination?

Unit 28

MIG SPOT WELDING

There are many welding applications that do not require **100** percent continuous welding of a particular joint. MIG spot welding is gaining wide acceptance today. It is competitive with riveting and resistance spot welding.

OPERATION

- Little welding skill is required.

- The operator places the gun nozzle against the metals to be joined and pulls the trigger.

- The welding control completes the welding cycle while the operator holds the torch in position.

- The preparation of the material depends on the strength required in the finished work.

APPLICATION

- Equipment should provide instantaneous and positive arc ignition, a constant rate of filler-wire feed, and precise timing of the weld cycle.

- An equipment setup includes constant-voltage power supply, and automatic spot-welding control, wire-feed unit and a gun and lead assumbly. A different type of nozzle than that used with normal MIG welding is required.

- MIG spot welding can be used on carbon steel, stainless steel, copper-bearing alloys, and all weldable aluminum alloys.

REVIEW QUESTIONS

1. What controls are different for MIG spot welding?

2. What kind of welding machine is used for MIG spot welding?

3. Does MIG spot welding require a lot of practice?

4. What one part on the MIG gun is different when spot welding?

5. What governs the strength of a weld?

ACKNOWLEDGMENTS

The authors wish to express their appreciation to the following for their assistance in the development of this text:

- Air Reduction Sales Company, New York, NY 10017
- Allegheny Ludlum Steel Corporation, Pittsburgh, PA 15222
- Aluminum Company of America, Pittsburgh, PA 15219
- American Welding Society, Miami, FL 33125
- Bausch and Lomb, Rochester, NY 14602
- Detroit Testing Machine Company, Detroit, MI 48213
- Dow Metal Products Company, Midland, MI 48640
- Hobart Brothers Company, Troy, OH 45373
- Larry Jeffus, technical photographer and welding consultant
- Lincoln Electric Company, Cleveland, OH 44117
- Linde Company, Division of Union Carbide Corp., New York, NY 10017
- Los Angeles Trade-Technical College
- Miller Electric Company, Appleton, WI 54911
- Niagara Mohawk Power Corp., Syracuse, NY 13202
- Norton Company, Worcester, MA 01616
- Revere Copper and Brass, Inc., New York, NY 10017
- Sylvania Electric Products Company, Towanda, PA 18848
- Tempil Corporation, New York, NY 10011
- Tweco Products, Inc., Wichita, KS 67277
- United States Steel Corporation, Pittsburgh, PA 15230
- Victor Equipment Company, Denton, TX 76203
- Welding Design and Fabrication, Cleveland, OH 44113
- Wilson Instrument, Division American Chain and Cable, New York, NY 10017